W9-AHP-559

Springer Series in

OPTICAL SCIENCES 96

founded by H.K.V. Lotsch

Springer Series in

OPTICAL SCIENCES

The Springer Series in Optical Sciences, under the leadership of Editor-in-Chief *William T. Rhodes*, Georgia Institute of Technology, USA, provides an expanding selection of research monographs in all major areas of optics: lasers and quantum optics, ultrafast phenomena, optical spectroscopy techniques, optoelectronics, quantum information, information optics, applied laser technology, industrial applications, and other topics of contemporary interest.
With this broad coverage of topics, the series is of use to all research scientists and engineers who need up-to-date reference books.

The editors encourage prospective authors to correspond with them in advance of submitting a manuscript. Submission of manuscripts should be made to the Editor-in-Chief or one of the Editors.

Motoichi Ohtsu (Ed.)

Progress in Nano-Electro-Optics III

Industrial Applications and Dynamics of the Nano-Optical System

With 186 Figures and 8 Tables

Professor Dr. Motoichi Ohtsu
The University of Tokyo
Department of Electronics Engineering
School of Engineering
7-3-1 Hongo, Bunkyo-ku
Tokyo 113-8656
Japan
E-mail: ohtsu@ee.t.u-tokyo.ac.jp

ISSN 0342-4111

ISBN 3-540-21050-4 Springer Berlin Heidelberg New York

Library of Congress Cataloging-in-Publication Data

Progress in nano-electro-optics III : industrial applications and dynamics of the nano-optical system / Motoichi Ohtsu (ed.). p.cm. – (Springer series in optical sciences ; v. 96)
Includes bibliographical references and index.
ISBN 3-540-21050-4 (alk. paper)
1. Electrooptics. 2. Nanotechnology. 3. Near-field microscopy. I. Ohtsu, Motoichi. II. Series.
TA1750 .P75 2002 621.381'045–dc21 2002030321

Springer is a part of Springer Science+Business Media

springeronline.com

Printed in Germany

Data prepared by the author using a Springer TEX macropackage
Data conversion by EDV-Beratung F. Herweg, Hirschberg
Cover concept by eStudio Calamar Steinen using a background picture from The Optics Project. Courtesy of John T. Foley, Professor, Department of Physics and Astronomy, Mississippi State University, USA.
Cover production: *design & production* GmbH, Heidelberg

Printed on acid-free paper SPIN 10985588 57/3141/di 5 4 3 2 1 0

Preface to *Progress in Nano-Electro-Optics*

Recent advances in electro-optical systems demand drastic increases in the degree of integration of photonic and electronic devices for large-capacity and ultrahigh-speed signal transmission and information processing. Device size has to be scaled down to nanometric dimensions to meet this requirement, which will become even stricter in the future. In the case of photonic devices, this requirement cannot be met only by decreasing the sizes of materials. It is indispensable to decrease the size of the electromagnetic field used as a carrier for signal transmission. Such a decrease in the size of the electromagnetic field beyond the diffraction limit of the propagating field can be realized in optical near fields.

Near-field optics has progressed rapidly in elucidating the science and technology of such fields. Exploiting an essential feature of optical near fields, i.e., the resonant interaction between electromagnetic fields and matter in nanometric regions, important applications and new directions such as studies in spatially resolved spectroscopy, nanofabrication, nanophotonic devices, ultrahigh-density optical memory, and atom manipulation have been realized and significant progress has been reported. Since nanotechnology for fabricating nanometric materials has progressed simultaneously, combining the products of these studies can open new fields to meet the above-described requirements of future technologies.

This unique monograph series entitled "Progress in Nano-Electro-Optics" is being introduced to review the results of advanced studies in the field of electro-optics at nanometric scales and covers the most recent topics of theoretical and experimental interest in relevant fields of study (e.g., classical and quantum optics, organic and inorganic material science and technology, surface science, spectroscopy, atom manipulation, photonics, and electronics). Each chapter is written by leading scientists in the relevant field. Thus, high-quality scientific and technical information is provided to scientists, engineers, and students who are and will be engaged in nano-electro-optics and nanophotonics research.

I gratefully thank the members of the editorial advisory board for valuable suggestions and comments on organizing this monograph series. I wish to express my special thanks to Dr. T. Asakura, Editor of the Springer Series in Optical Sciences, Professor Emeritus, Hokkaido University for recommending me to publish this monograph series. Finally, I extend an acknowledgement to Dr. Claus Ascheron of Springer-Verlag, for his guidance and suggestions, and to Dr. H. Ito, an associate editor, for his assistance throughout the preparation of this monograph series.

Yokohama, October 2002 *Motoichi Ohtsu*

Preface to Volume III

This volume contains five review articles focusing on different aspects of nanoelectro-optics. The first article reviews fabrications of advanced fiber probes, i.e., application-oriented fiber probes, and applications to microscopy and spectroscopy.

The second article is devoted to reviewing a localized surface plasmon sensor using uniform surface-adsorbed metal particles. The unique feature of this sensor is exploited to make contributions to the field of life sciences, culminating in productive interactions between nano-optics and biotechnology.

The third article concerns industrial applications to ultrahigh-density optical storage. A highly efficient near-field optical head, using a wedge-shaped metallic plate, is described.

The fourth article also deals with ultrahigh-density optical storage. Circumferential magnetic patterned media are reviewed, which are prepared by an artificially assisted self-assembling method. These media can be used to realize a novel storage system with high contrast and high signal-to-noise ratio.

The last article reviews the study on quantum optical near-field interaction of nanometric particles, i.e., transfer of the dipole moments of the systems and dynamics of localized photons in an open system. Through this study, two kinds of phases are predicted: storage and nonstorage modes of localized photons in the system. Finally, chaotic behavior, due to the nonlinearity of the equations of motion, is described.

As was the case of Volumes I and II, this volume is published with the support of an associate editor and members of editorial advisory board. They are:

I hope that this volume will be a valuable resource for the readers and future specialists.

Yokohama, July 2004 *Motoichi Ohtsu*

Contents

List of Contributors

Koji Asakawa
Advanced Materials
and Devices Laboratory
Corporate Research
and Development Center
Toshiba Corporation
1 Komukai-Toshiba-cho, Saiwai-ku
Kawasaki 212-8582, Japan
kouji.asakawa@toshiba.co.jp

Hiroyuki Hieda
Storage Materials
and Devices Laboratory
Corporate Research
and Development Center
Toshiba Corporation
1 Komukai-Toshiba-cho, Saiwai-ku
Kawasaki 212-8582, Japan
hiroyuki.hieda@toshiba.co.jp

Michael Himmelhaus
Angewandte Physikalische Chemie
Universitaet Heidelberg
Im Neuenheimer Feld 253
D-69120 Heidelberg Germany
i17@ix.urz.uni-heidelberg.de

Takashi Ishino
Storage Materials
and Devices Laboratory
Corporate Research
and Development Center
Toshiba Corporation
1 Komukai-Toshiba-cho, Saiwai-ku
Kawasaki 212-8582, Japan
takashi.ishino@toshiba.co.jp

Yoshiyuki Kamata
Storage Materials
and Devices Laboratory
Corporate Research
and Development Center
Toshiba Corporation
1 Komukai-Toshiba-cho, Saiwai-ku
Kawasaki 212-8582, Japan
yoshiyuki.kamata@toshiba.co.jp

Akira Kikitsu
Storage Materials
and Devices Laboratory
Corporate Research
and Development Center
Toshiba Corporation
1 Komukai-Toshiba-cho, Saiwai-ku
Kawasaki 212-8582, Japan
akira.kikitsu@toshiba.co.jp

Kazuo Kitahara
Division of Natural Sciences
International Christian University
Tokyo 181-8585, Japan
kazuo@icu.ac.jp

Kiyoshi Kobayashi
ERATO Localized Photon Project
Japan Science
and Technology Corporation
687-1 Tsuruma Machida
Tokyo 194-0004, Japan
kkoba@ohtsu.jst.go.jp

Takuya Matsumoto
Storage Technology Research Center
Research and Development Group
Hitachi, Ltd.
1-280 Higashi-Koigakubo, Kokubunji
Tokyo 185-8601, Japan
m-takuya@rd.hitachi.co.jp

Shuji Mononobe
PRESTO, Japan Science
and Technology Agency
Kanagawa Academy of Science
and Technology
KSP East 408, 3-2-1 Sakado,
Takatsu-ku
Kawasaki 213-0012, Japan
mononobe@net.ksp.or.jp

Seiji Morita
Storage Materials
and Devices Laboratory
Corporate Research
and Development Center
Toshiba Corporation
1 Komukai-Toshiba-cho, Saiwai-ku
Kawasaki 212-8582, Japan
sj.morita@toshiba.co.jp

Katsuyuki Naito
Storage Materials
and Devices Laboratory
Corporate Research
and Development Center
Toshiba Corporation
1 Komukai-Toshiba-cho, Saiwai-ku
Kawasaki 212-8582, Japan
katsuyuki.naito@toshiba.co.jp

Motoichi Ohtsu
Interdisciplinary Graduate School
of Science and Technology
Tokyo Institute of Technology
4259 Nagatsuta-cho, Midori-ku
Yokohama 226-8502, Japan
ohtsu@ae.titech.ac.jp

Masatoshi Sakurai
Storage Materials
and Devices Laboratory
Corporate Research
and Development Center
Toshiba Corporation
1 Komukai-Toshiba-cho, Saiwai-ku
Kawasaki 212-8582, Japan
masatoshi.sakurai@toshiba.co.jp

Suguru Sangu
ERATO Localized Photon Project
Japan Science
and Technology Corporation
687-1 Tsuruma Machida
Tokyo 194-0004, Japan
sangu@ohtsu.jst.go.jp

Akira Shojiguchi
ERATO Localized Photon Project
Japan Science
and Technology Corporation
687-1 Tsuruma Machida
Tokyo 194-0004, Japan
akirasho@ohtsu.jst.go.jp

Hiroyuki Takei
Mechanical Engineering Research
Laboratory
Hitachi, Ltd.
502 Kandatsu, Tsuchiura
Ibaraki 300-0013, Japan
htakei@gm.merl.hitachi.co.jp

Kuniyoshi Tanaka
Storage Materials
and Devices Laboratory
Corporate Research
and Development Center
Toshiba Corporation
1 Komukai-Toshiba-cho, Saiwai-ku
Kawasaki 212-8582, Japan
kuniyoshi.tanaka@toshiba.co.jp

Near-Field Optical Fiber Probes and the Imaging Applications

S. Mononobe

1 Introduction

Recently, scanning near-field optical microscopy (SNOM) employing an apertured probe has been seen as a powerful tool for nano-optical applications due to its high spatial resolution capability down to a few tens of nanometers or less. The resolution is mainly determined by the aperture size of the probe and the sample–probe distance because the SNOM is a scanning probe microscope based on the short-range electromagnetic interaction between the probe and sample, which are much smaller than the optical wavelength. It is apparent that fabricating such probes and controlling the sample–probe distance have been the most important factors in the development of high-resolution SNOM.

A near-field optical interaction system can be operated in two complementary ways called illumination mode and collection mode, where the probe acts as a selective generator of a localized optical near-field and a sensitive scatterer of an optical near-field generated over the sample, respectively. Figures 1a and b show schematic illustrations of i-mode SNOM and c-mode SNOM, respectively. In the c-mode SNOM where the light is incident to the total internal reflection, the sample–probe distance can be controlled by using near-field optical (NFO) variation as a feedback control signal. For i-mode SNOM, a shear-force feedback technique [1,2] has been widely used to regulate the sample–probe distance.

In order to realize such a scattering probe and a generating probe, forming a dielectric taper and metallizing the taper has been used as an effective method. Tapered probes have been fabricated by the following methods:

- Etching a quartz crystal rod [3]
- Heating and pulling a glass pipette by a micropipette puller [4]
- Meniscus etching of an optical fiber [5–13]
- Heating and pulling an optical fiber [14–18]
- Selective etching of an optical fiber [19–23]
- Heating and pulling and etching of an optical fiber [24,25]
- Meniscus etching and selective etching of an optical fiber [26,27]

For metallizing the tapered probes, a vacuum evaporation method [28] has been applied, where the dielectric taper is coated with an aluminum or gold

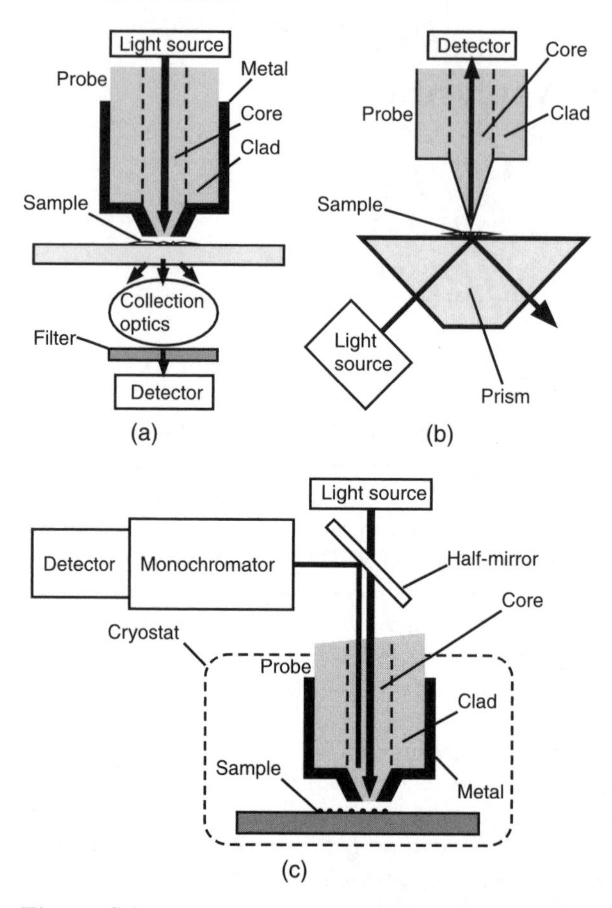

Fig. 1. Schematic illustrations of (**a**) the i-mode, (**b**) the c-mode, and (**c**) the i-c mode. In the c-mode SNOM, the light is incident to the total internal reflection. The three-dimensional optical near field generated and localized on the sample surface is scattered by the probe, and part of the scattered field is collected and detected through the probe. The principle of operation of i-mode is similar except that the probe works as a generator of the optical near field to illuminate the sample surface. The scattered field by the sample is collected by conventional optics. In i-c mode, a sample is excited by an optical near field on the probe. The light generated on the sample is scattered and collected by the probe

film by a vacuum evaporation unit except for its apex region. Among the metallized tapered probes, fiber-type probes have high transmission efficiencies due to their waveguide structure, and moreover, have been demonstrated to have the molecular sensitivity, the nanometric spatial resolution, and the locally spectroscopic capability in i-mode SNOM, c-mode SNOM, and hybrid SNOM called illumination-collection mode (i-c mode). Figure 1c shows schematic illustration of i-c mode SNOM where the apertured probe functions as both an optical generator exciting the sample and a scatterer of the excited optical near-fields.

i-mode SNOM and i-c mode SNOM are often employed for near-field imaging of dye-doped samples and spectroscopic investigation of semiconductor devices. In such spectroscopic applications where one must cope with extremely low detected power, the probe should have high throughput in i-mode SNOM, and be highly sensitive in c-mode SNOM. Furthermore, in the i-mode fashion, to avoid thermal damage to the sample and the probe, the probe should be used with as low an input power as possible. Therefore, the resolution capability and throughput of the tapered probe have to be optimized depending on SNOM applications. This optimization should be done by varying the probe shape, i.e., the cone angle. Furthermore, to obtain a highly resolved image, one must fabricate a metallized probe with an apex region emerging from a metal film.

We have developed several application-oriented fiber probes called protrusion-type probe, double-tapered probe, pure silica fiber probe, and performed near-field imaging of biomolecules and dye-doped samples, spectroscopic study of semiconductors, and ultraviolet applications. In this chapter, the near-field optical fiber probes and near-field microscopy applications are described. Section 2 discusses the basic techniques for tapering and metallizing fibers. The protrusion-type probe and the biological imaging are described in Sect. 3. Section 4 describes apertured probes and their application to near-field imaging of dye-doped samples. The double-tapered probe and pure silica fiber probe are described in Sects. 5 and 6, respectively. Table 1 summarizes the SNOM images appeared in this chapter.

2 Basic Techniques for Tapering and Metallizing Optical Fibers

For tapering an optical fiber, three basic techniques, i.e., heating and pulling, meniscus etching, and selective etching have been used. The characteristics of these techniques are summarized in Table 2.

2.1 Heating-and-pulling and Metallization Techniques

In the heating-and-pulling technique [14], a silica-based optical fiber is heated and pulled by a micropipette puller combined with a CO_2-gas laser as shown in Fig. 2. One can fabricate a tapered fiber with an apex diameter of 50 nm and a cone angle of 20–40° by a commercial micropipette puller. This tapering can be applied to any optical fiber with a diameter of more than 125 μm by adjusting the laser power, the strength of the pull, and the delay time between the end of the heating and the beginning of the pulling. However, it is difficult to control the cone angle while maintaining an apex diameter

Table 1. List of the SNOM images appeared in this chapter

Fig.	Image	Probe type	Mode	Feedback	Ref.
15c	Near-field image of salmonella flagellar filaments in air	Protrusion	c-mode	NFO intensity	[29]
15d	Near-field image of salmonella flagellar filaments in water	Protrusion	c-mode	NFO intensity	[30]
18c	Near-field image microtubules of a pig brain	Protrusion	c-mode	Shear-force	[31]
20d–f	Photoluminescence images of GaAs quantum dots	Protrusion	i-mode	Shear-force	[32]
24b	Near-field image of deoxyribo-nucleic acid	Ag-MgF_2-Al-coated	c-mode	NFO intensity	[33]
26b, c	Near-field images of neurons and microtubules labeled with toluidine blue	Apertured	i-mode	Shear-force	[34]
27b	Fluorescence image of Rhodamin 6G molecules	Double-tapered	i-c mode	Shear-force	[35]
31b	Photoluminescence imaging lateral p-n junction based on GaAs	Double-tapered	i-c mode	Shear-force	[36]
32b	Raman spectroscopic mapping of tabular polydyacetylene single crystal	Double-tapered	i-mode	Shear-force	[37]
40b	UV photoluminescence image of polydihexylsilane	Triple-tapered with a pure silica core	i-mode	Shear-force	[23]
41b	UV photoluminescence image of n-decyl-(s)-2-methylbutyl silane	Triple-tapered with a pure silica core	i-mode	Shear-force	[38]
45	UV photoluminescence image of polydihexylsilane	Pulled and etched with a pure silica core	i-mode	Shear-force	[39]

Table 2. Characteristics of the three basic techniques for tapering an optical fiber

Technique	Cone angle θ	Apex diameter d	Reproducibility
Meniscus etching	9–40°	60 nm or more	80% or less
Selective etching	14–180°	10 nm	Almost 100%
Pulling	20–40°	50 nm	Around 80%

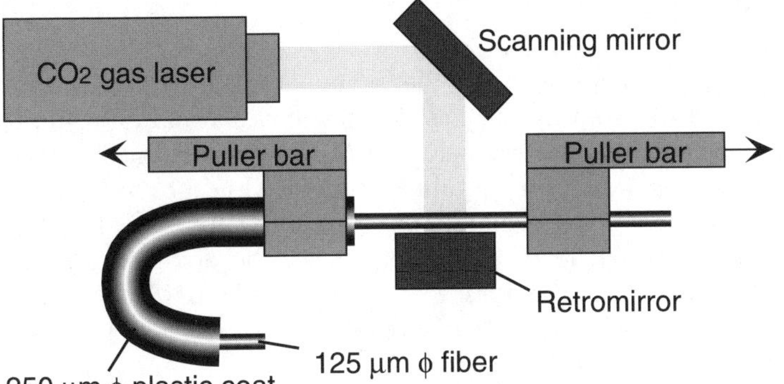

Fig. 2. Schematic illustration of the micropipette puller used for tapering an optical fiber with a diameter of 125 μm

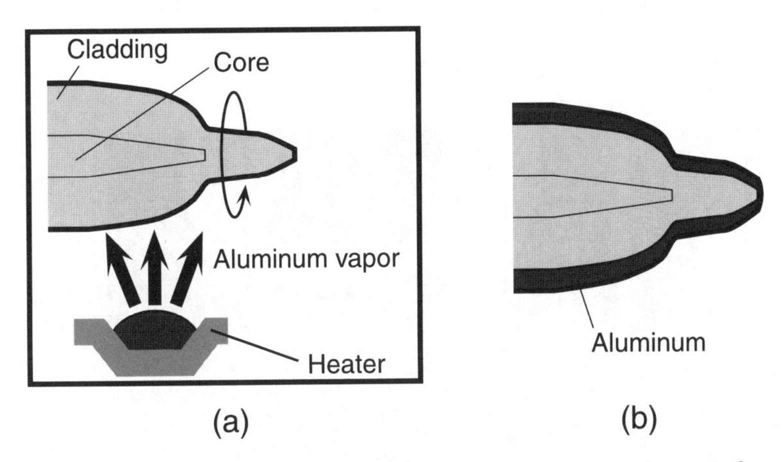

Fig. 3. Schematic illustrations of (**a**) the vacuum evaporation unit for metallizing the pulled fiber and (**b**) the metallized fiber

as small as 50 nm. In the tapered portion, strong optical leaky modes are generated due to the varied core diameter.

For i-mode SNOM, the pulled fiber with an apex diameter of about 50 nm must be metallized except for its apex region. To metallize the fiber, the pulled fiber is rotated while evaporating aluminum in vacuum as shown in Fig. 3a so that the metallized probe has a thickness profile as shown in Fig. 3b. Here, the typical radial thickness is around 150 nm. The apex region is aluminized with a thickness smaller than the half-radial thickness due to the throwing of evaporated vapor. The metal thickness covering the apex region can be reduced to a quarter of the radial thickness by inclining the rotating fiber in vacuum.

2.2 Meniscus Etching

In the meniscus etching[1], a single-mode fiber is immersed in HF acid with a surface layer of an organic solution such as silicone oil as shown in Fig. 4a. It is tapered with a cone angle since the height of the meniscus formed around the fiber is reduced depending on the fiber diameter (Fig. 4b). When the fiber is completely tapered, the etching stops automatically (Fig. 4c). The cone angle can be increased up to 35–40°. However, the obtained tapered fiber has a geometrically eccentric apex with an elliptical cross section. The longer and shorter principal diameters of this elliptical apex take values of 200 nm and 10–20 nm, respectively.

The vacuum evaporation method as shown in Fig. 3a can be applied to metallize the meniscus-etched fibers. Prior to metallizing, one has to remove organic dirt from the surface of the etched fibers by cleaning techniques, i.e., immersing in a H_2SO_4 solution.

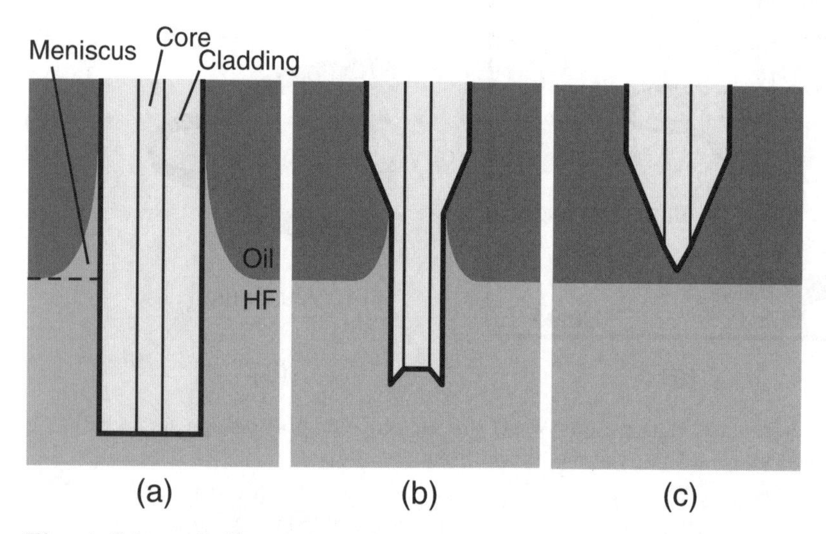

Fig. 4. Schematic illustrations of meniscus etching of a fiber at (**a**) the start, (**b**) tapering, and (**c**) stop

2.3 Selective Etching

By applying the selective-etching method to a highly GeO_2-doped fiber, one can obtain probe tips with a small apex diameter less than 10 nm. By varying the concentration of etching solutions based on hydrogen fluoride (HF) and

[1] The meniscus-etching technique was originally developed to fabricate a fiber-optic microlens.

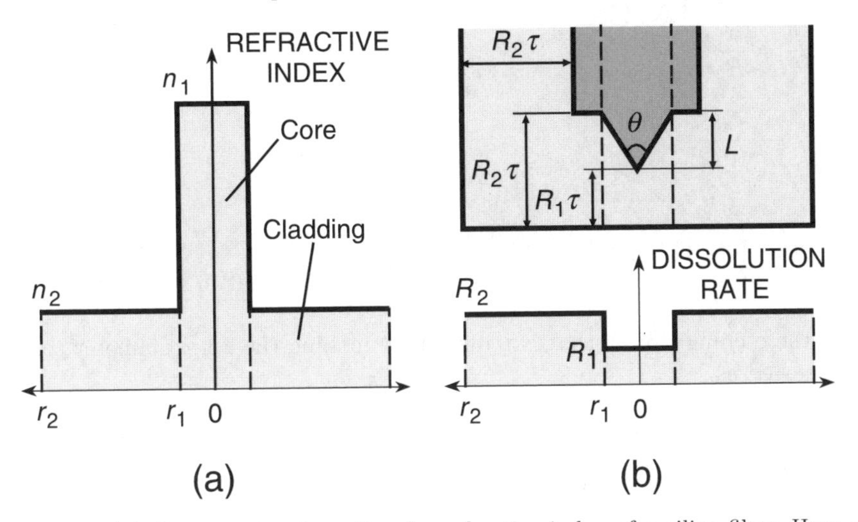

Fig. 5. (**a**) Cross-sectional profile of a refractive index of a silica fiber. Here, n_1 and n_2 are the refractive indices of the core and clad, respectively; r_1 and r_2 are the radii of the core and clad, respectively. (**b**) *Top*, a geometrical model for the tapering process; Here, τ is the etching time required for making the apex diameter zero. θ is the cone angle of the tapered core. L is the length of the tapered core. *Bottom*, cross-sectional profiles of the dissolution rates R_1 and R_2 of the core and clad

ammonium fluoride (NH_4F), the cone angle can be controlled in a wide region from 20° to 180° for an apex diameter less than 10 nm. Furthermore, selective etching is the most highly reproducible technique among the three tapering techniques. This method can be applied to any single-mode fiber produced by vapor-phase axial deposition (VAD) [40].

Figure 5a shows a cross-sectional profile of the refractive index of a silica fiber with a GeO_2-doped core and a pure silica clad. Here, n_1 and n_2 are the refractive indexes of the core and clad, respectively. r_1 and r_2 are the radii of the core and clad, respectively. On immersing the fiber in a buffered HF solution (BHF) with a volume ratio of [40%–NH_4F aqueous sol.]: [50%–HF acid]: [H_2O]= X:1:1 at 25°C, the core is hollowed at $X = 0$ and is tapered at $X = 10$. Figure 5b shows a schematic explanation of the geometrical model for the tapering process based on selective etching. Bright shading and dark shading in the upper part represent the cross-sectional profiles of the fiber before and after the etching with an etching time τ, respectively. τ is the etching time required for making the apex diameter zero, θ and d are the cone angle and apex diameter of the probe, respectively. The lower part shows the dissolution rates R_1 and R_2 of the core and clad, respectively. Here, $R_1 < R_2$. Assuming that the dissolution rates R_1 and R_2 are constant within the core and clad regions, respectively, relations between the cone angle θ, the length L of the tapered core, and the apex diameter d are represented by

$$\sin\frac{\theta}{2} = \frac{R_1}{R_2}\,, \tag{1}$$

$$L = \frac{r_1 - d/2}{\tan(\theta/2)}\,, \tag{2}$$

and

$$d(T) = \begin{cases} 2r_1(1 - T/\tau) & (T < \tau)\,, \\ 0 & (T \geq \tau)\,. \end{cases} \tag{3}$$

Here, the etching time τ that is required for making the apex diameter zero is expressed as

$$\tau = \frac{r_1}{R_1}\sqrt{\frac{R_1 + R_2}{R_2 - R_1}}\,. \tag{4}$$

GeO_2-Doped Fiber

When a fiber with a GeO_2-doped core and a pure silica clad is immersed in BHFs with volume ratios of X:1:1, the core region is hollowed in $X < 1.7$, and is tapered in $X > 1.7$ as shown in Fig. 5. Based on (1), the cone angle is determined by the dissolution rate ratio of the core and clad. Figure 6a shows variations of the dissolution rates R_1 and R_2 and the cone angle θ as a function of X. Here, the fiber used was produced by vapor-phase axial deposition (VAD) so as to have an index difference of 2.5%. The dissolution rate ratio R_1/R_2 decreases with increasing NH_4F volume ratio of X, and

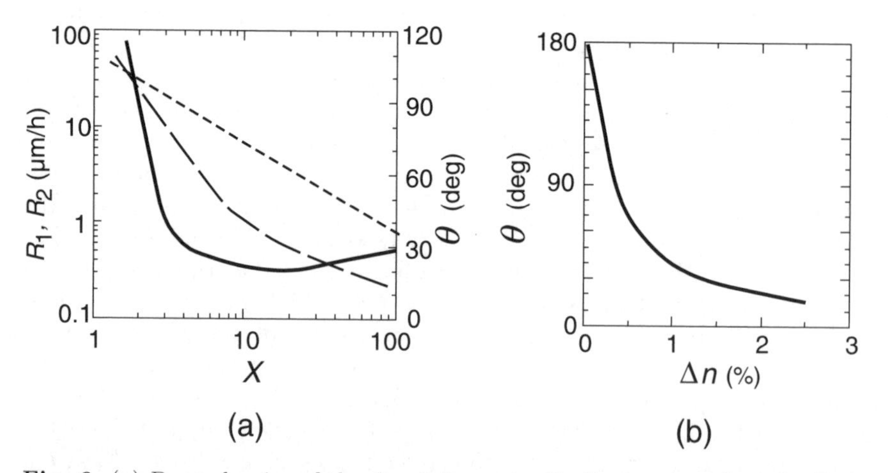

Fig. 6. (**a**) Dependencies of the dissolution rate R_1 (*broken curve*) of the GeO_2-doped core, R_2 (*dotted line*) of the pure silica clad, and the cone angle θ (*solid curve*) on X. Here, the value of Δn defined as $(n_1^2 - n_2^2)/2n_1^2$ is 2.5%; (**b**) Dependency of θ on Δn at $X = 10$

approaches a constant value at $X = 10$–30. The cone angle that is determined by the ratio R_1/R_2 takes a minimum value of 20° at $X = 10$. When X is fixed, the cone angle is determined by the index difference. Figure 6b shows the dependence of the cone angle on the index difference at $X = 10$. The index difference is increased by increasing the GeO_2 doping ratio.

The VAD is an effective method to produce a GeO_2-doped fiber with an index difference as large as 2.5–3.0%. A dispersion-compensating fiber (DCF) with an index difference of $\Delta n = 2.5\%$ and core diameter of 2 μm can be applied to various selective etching methods of fabricating a protrusion-type probe, a double-tapered probe, etc. This DCF was originally produced as a device for controlling the optical dispersion of a 1500-nm optical transmission system by VAD, and has a cutoff wavelength of around 0.8 μm.

By immersing the DCF in BHF with a volume ratio of 10:1:1 at 25°C, a tapered fiber probe with a small cone angle of 20° and an apex diameter less than 10 nm is fabricated with almost 100% reproducibility. Furthermore, the cone angle can be controlled as $20° \leq \theta < 180°$ by varying the volume ratio X of BHF, as shown in Fig. 6a. Such high controllability of the cone angle is indispensable for tailoring a high-throughput probe and a high-resolution probe.

Pure Silica Core Fiber with a Fluorine-Doped Clad

Pure silica core fiber has high transmittance in a wide region from near ultraviolet to near infrared (see Fig. 33). To make an index difference between the pure silica core and the clad, the clad is often made of fluorine-doped silica. We produced a pure silica core fiber with the fluorine-doped clad and an index difference −0.7%.[2] By immersing the fiber in BHFs with volume ratios 1.7:1:1 and 10:1:1, the obtained probe has cone angles of 62° and 77°, respectively. The fiber could not be hollowed at any volume ratios of BHFs. In selective etching of a pure silica fiber, varying the index difference or the fluorine-doping ratio controls the cone angle. Among single-mode fibers, only 1500-nm pure silica fiber (PSF) with an index difference of −0.3% and core diameter of 10 μm is commercially available.

Controlling the Clad Diameter

For the shear-force feedback technique, the clad diameter of the probe is one of the main parameters governing the resonance frequency of dithering a probe. In our experience, tapered probes with a clad diameter of 20–30 μm are usually needed to obtain a proper resonance frequency, i.e., 20–40 kHz. Figure 7 schematically shows an etching method to fabricate tapered probes

[2] For the following discussion, we define the relative refractive index difference Δn of doped glass to pure silica, which is expressed as $(n_2^2 - n_1^2)/2n_2^2$ and $(n_1^2 - n_2^2)/2n_1^2$ for a pure silica core fiber and a pure silica clad fiber, respectively.

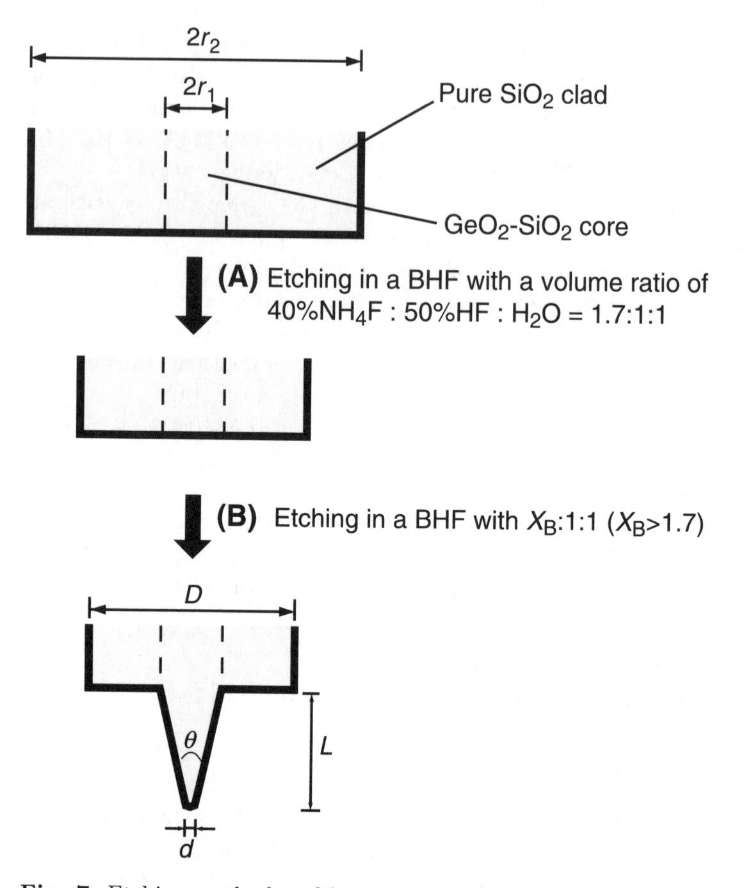

Fig. 7. Etching method to fabricate a shoulder-shaped probe. r_1, r_2, radii of the core and clad, respectively. D, reduced clad diameter. θ, cone angle of the tapered core. L, length of the tapered core

with the reduced clad diameter. Here, r_2 is the clad radius before etching, D is the clad diameter of the probe, θ is the cone angle of the tapered core, d is the apex diameter, and L is the length of the tapered core. This method involves two steps: (A) reducing the clad thickness and (B) tapering the core.

In step A, the fiber is immersed in BHF with volume ratio of [NH_4F aqueous solution (40%)]: [HF acid (50%)]: [H_2O]=1.7:1:1 for an etching time T_A. Then, the clad diameter is reduced to $[2r_2 - 2R_{2A}T_A]$. The core end is kept flat since the dissolution rates R_{1A} and R_{2A} of the core and clad are equal.

In step B, the fiber is selectively etched in X_B:1:1 (where $X_B > 1.7$) for etching time T_B. The core is tapered with the cone angle θ represented by

$$\sin\frac{\theta}{2} = \frac{R_{1B}}{R_{2B}} . \tag{5}$$

Fig. 8. SEM micrographs of (**a**) a shoulder-shaped probe and (**b**) the magnified apex region. $D = 25$ μm; $\theta = 20°$; $d < 10$ nm

To make the apex diameter zero, the etching time T_{B} must be longer than the time τ, which is given by

$$\tau = \frac{r_1}{R_{1\mathrm{B}}}\sqrt{\frac{R_{1\mathrm{B}} + R_{2\mathrm{B}}}{R_{2\mathrm{B}} - R_{1\mathrm{B}}}} . \tag{6}$$

The clad diameter D is represented by

$$D = 2r_2 - 2(R_{2\mathrm{A}}T_{\mathrm{A}} + R_{2\mathrm{B}}T_{\mathrm{B}}) , \tag{7}$$

which is proportional to the etching time T_{B}.

Using a DCF with a core diameter of 2 μm, a clad diameter of 125 μm, and an index difference of 2.5%, we obtained a tapered probe with a cone angle of $\theta = 20°$ and a clad diameter of $D = 25$ μm. Figures 8a and 8b show SEM micrographs of the probe and the magnified top region, respectively. The conditions of the etching process are summarized in Table 3. By substituting $r_1 = 1$ μm, $R_{1\mathrm{B}} = 1.1$ μm h^{-1}, and $R_{2\mathrm{B}} = 6.5$ μm h^{-1} into (6), an estimated value of $\tau = 64$ min is obtained. For $T_{\mathrm{B}} = 75$ min ($> \tau$), the clad diameter D is represented by $D = -60T_{\mathrm{A}} + 106$ [μm]. The cone angle is controlled by varying the NH_4F volume ratio in step B. For the dependence of the cone angle on the volume ratio, refer to Fig. 6a.

Table 3. Conditions for fabricating the probe as shown in Fig. 8a

Step	Volume ratio of BHF	Etching time (min)	Dissolution rates (μm h^{-1})
A	1.7:1:1	86	$R_1 = R_2 = 30$
B	10:1:1	75	$R_1 = 1.1$, $R_2 = 6.5$

3 Protrusion-type Probe and Its Imaging Applications

3.1 Protrusion-type Probe

The uncoated fiber probe with a nanometric apex can be employed for the c-mode SNOM as shown in Fig. 1b. Although the principal factor governing the resolution is the small apex size, the resolution is affected by propagating components and the scattering of low spatial frequency components of the optical near field. To obtain highly resolved c-mode images, such unwanted scattered light must be suppressed. To perform high-resolution near-field imaging, we developed a fiber probe with a protruding tip from a metal film, which is called a protrusion-type probe [41]. Figure 9a shows the cross-sectional schematic illustration of the protrusion-type probe. Here, θ, d, d_f, and t_M represent the cone angle, the apex diameter, the foot diameter, and the thickness of the metal film, respectively. This probe scatters frequency components between $(1/d_f)$ and $(1/d)$ and suppresses the generation of components of less than $(1/d_f)$. To realize such a probe, a selective resin-coating (SRC) method [41] was developed, which can be applied to tapered probes fabricated by the etching method as shown in Fig. 7. Apertured probes are also effective for suppressing unwanted light. However, it is difficult to produce apertured probes with an apex diameter of a few tens of nanometers by a vacuum evaporation unit. If applying the vacuum evaporation method to a fiber tip with a cone angle of 20°, an entirely coated fiber probe as shown in Fig. 9b is obtained. The thickness t_A of metal covering the apex could not be made less than 40 nm for $t_M = 150$ nm.

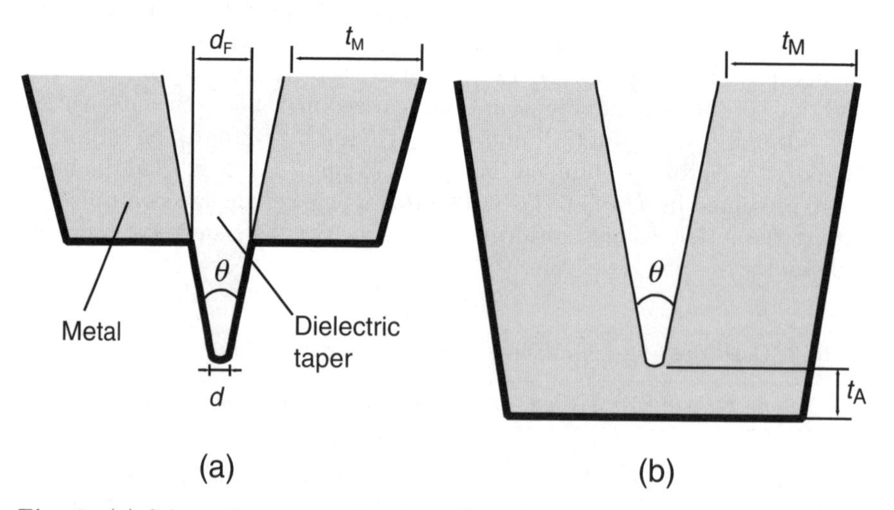

Fig. 9. (**a**) Schematic cross-sectional profiles of a protrusion-type probe and (**b**) entirely coated probe. θ, cone angle of the taper; d, d_f, apex and foot diameters of the protruding tip, respectively; t_M, thickness of the metallic film; t_A, thickness of metal covering the apex

3.2 Fabricating Protrusion-type Probes by Selective Resin-Coating Method

Figure 10 shows a schematic diagram of the SRC involving four steps: (A) metal coating, (B) resin coating, (C) preferential etching of metal covering the apex region, (D) removing the resin. Figure 11 shows a scanning electron micrograph of the top region of the fabricated protrusion-type probe. The probe has a protrusion from a gold film and a foot diameter d_f of less than 30 nm.

Before applying the SRC method, a tapered probe was fabricated with a clad diameter $D = 45\ \mu m$ and a cone angle $\theta = 20°$ based on selective etching. In step A, the probe was coated with 120-nm thick gold by a magnetron sputtering unit. In step B, the probe is dipped in an acrylic resin solution and then removed from the acrylic solution with a withdrawal speed of $V_D = 5\ cm\ s^{-1}$. Using an acrylic resin solution with a low viscosity coefficient of 11 cP, we succeed in making a fiber tip protruding from resin film. The density of the acrylic solution is $0.85\ g\ cm^{-3}$ at 25°C. In step C, the probe was etched for 2 min in a solution $KI\text{-}I_2\text{-}H_2O$, mixed with a weight ratio of 20:1:400 and diluted 50 times with deionized water. In step D, the acrylic film was removed by immersing the probe in acetone.

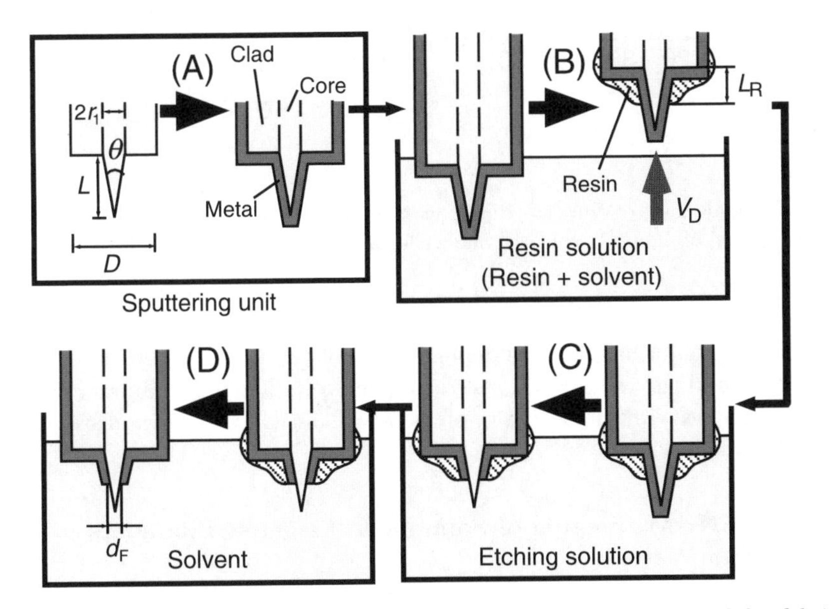

Fig. 10. Schematic diagram of the selective resin-coating method for fabricating a protrusion-type probe; D, clad diameter of the probe; r_1, core radius; θ, L, cone angle and length of the tapered core, respectively; V_D, withdrawal speed of the fiber from the resin solution; L_R, length of the tapered core where the resin is coated; d_f, foot diameter of the protrusion

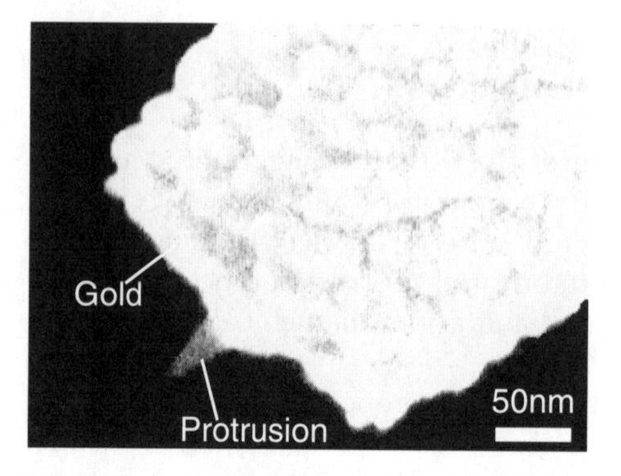

Fig. 11. Scanning electron micrograph of a protrusion-type probe fabricated by the SRC method as shown in Fig. 10; $\theta = 20°$; $d < 10$ nm; $d_f = 30$ nm

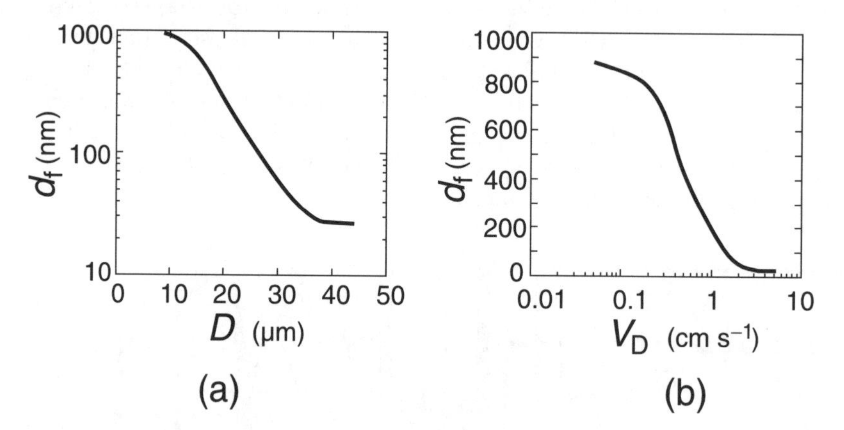

Fig. 12. Dependencies of the foot diameter d_f on (**a**) the clad diameter D for a withdrawal speed of 5 cm s^{-1} and (**b**) the withdrawal speed V_D for a clad diameter of 45 μm

Figures 12a and 12b show the dependencies of the foot diameter on the clad diameter and the withdrawal speed, respectively. The foot diameter can be controlled by varying the viscosity of the resin solution, withdrawal speed, and the clad diameter.

3.3 c-mode SNOM Imaging of Salmonella Flagellar Filaments in Air and Water

Using a protrusion-type probe fabricated by selective etching and selective resin coating, Naya et al. [29, 30] succeeded in obtaining high-resolution c-mode SNOM images of Salmonella flagellar filaments in both air and water.

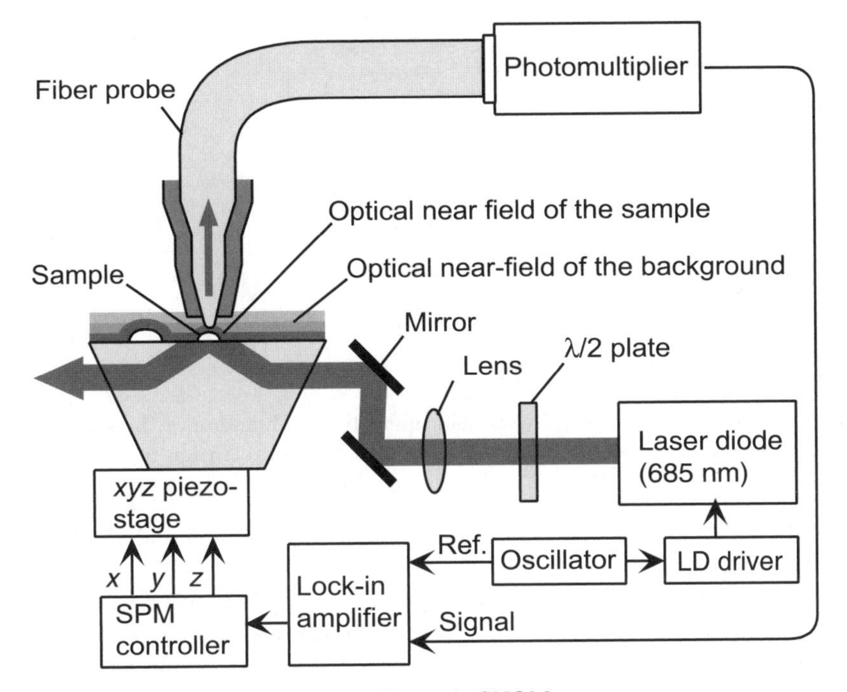

Fig. 13. Schematic illustration of c-mode SNOM

Instrument of the c-mode SNOM

Figure 13 shows a schematic illustration of the optical system of c-mode SNOM under near-field optical intensity feedback control. The sample substrate is mounted on the xyz piezo-stage. The piezo parameters are chosen depending on features such as the lateral and longitudinal variations of the sample. Here, the piezo-stage has a total lateral span of 100 µm × 100 µm. The z stage has a span of 5 µm. A right-angle prism or dove prism is used for generating the optical near field on the sample surface. Here, care should be taken to avoid unwanted scattering light generated from the prism surface due to the dust and scratches. The probe is mounted on a xyz stage.

The light from a laser diode of 680 nm wavelength was used as the light source. The light is incident on the prism and is totally internally reflected. Then, the optical near field is generated on the sample surface. As shown by the shaded region in Fig. 13, the optical near field is localized three-dimensionally, and follows the spatial variations of the sample. When the probe approaches closer to the sample surface, the optical near field is detected through the probe by a highly sensitive detector such as a photomultiplier tube. By scanning the sample, information about the sample feature can be extracted. A typical value of the detected power is of the order of 1 nW to 100 pW. In this system, the signal over the background can be effectively

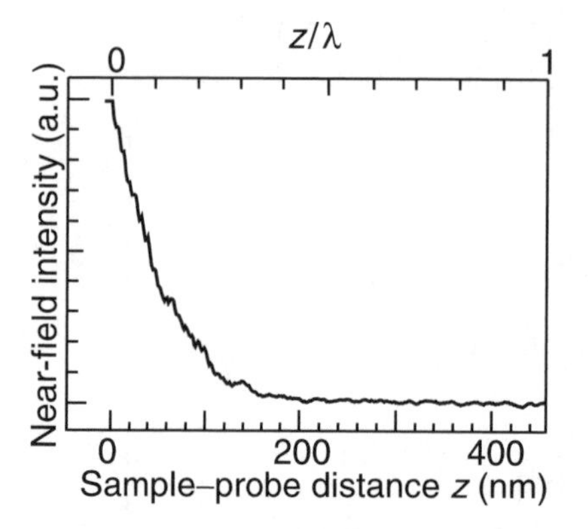

Fig. 14. Variation of the optical near-field intensity as a function of the sample–probe distance, obtained with c-mode SNOM employing a protrusion-type probe

extracted by modulating the light and by amplifying the modulated signal with a lock-in amplifier.

Figure 14 shows the variation of the optical near-field intensity as a function of the sample–probe distance, which is obtained with c-mode SNOM employing a protrusion-type probe with a foot diameter of 30 nm as shown in Fig. 15a. The optical near-field intensity sharply increases by decreasing the sample–probe distance. This sharp variation of the near-field intensity can be used as a feedback control signal to regulate the sample–probe distance.

Sample

The samples were fabricated by fixing the separated filaments from the body on a hydrophilic glass substrate. A single filament has a typical length of a few micrometers and a diameter of around 25 nm as shown in Fig. 15b. The sample substrate was mounted on the prism of the optical system as Fig. 13.

Near-Field Optical Images of Salmonella Flagellar Filaments

Figures 15c shows a near-field image of Salmonella flagellar filaments, obtained by the c-mode SNOM under near-field optical (NFO) intensity feedback control. Here, the incident light was s-polarized. The pixel size is 25 nm $\times$ 25 nm. The sample–probe distance was kept at less than 15 nm. The full width at half-maximum (FWHM) of the region as indicated by arrows is 50 nm. This value is comparable to the diameter obtained from the TEM image (Fig. 15a), and indicates the high-resolution capability of SNOM employing the protrusion-type probe.

Figure 15d shows the near-field optical image of Salmonella flagellar filaments obtained in water. The pixel size is 10 nm $\times$ 10 nm. This near-field

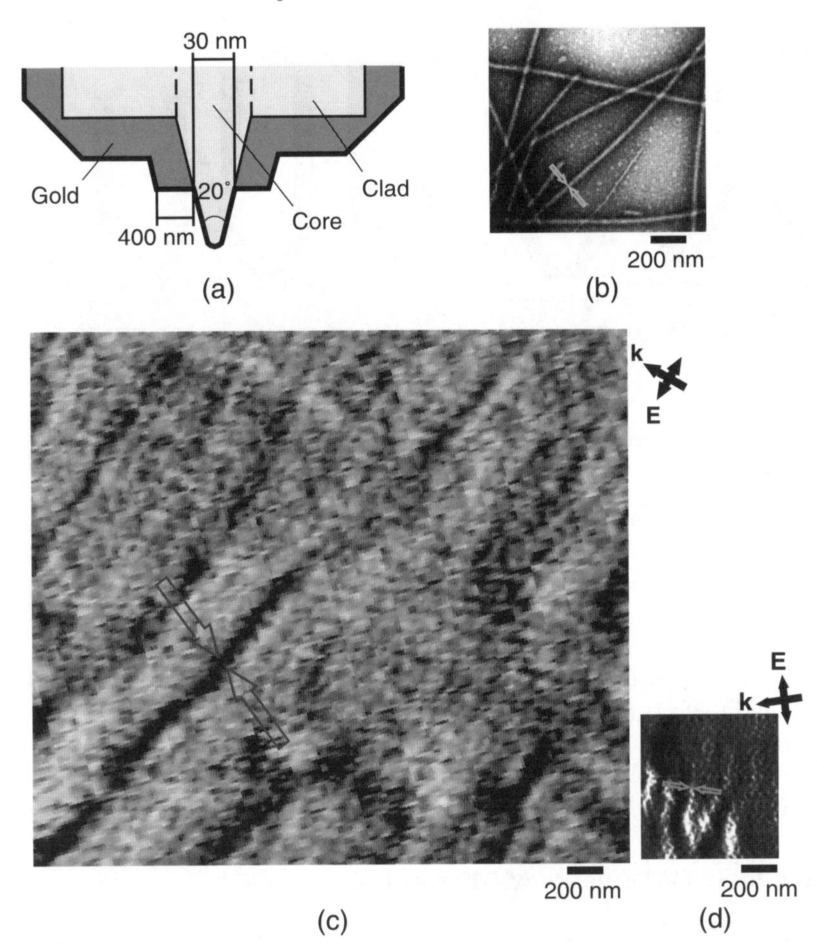

Fig. 15. (**a**) Schematic illustration of the fabricated protrusion-type fiber probe with cone angle of 20°, a foot diameter of 30 nm, and apex diameter less than 10 nm; (**b**) A transmission electron micrograph of the Salmonella flagellar filaments. The diameter of a flagellar filament, as indicated by arrows, are 25 nm; (**c**) A c-mode SNOM image of Salmonella flagellar filaments. The arrows at the *top right-hand side* represent the directions of wavevector **k** and electric field vector **E**; (**d**) Perspective view of the near-field image of Salmonella flagellar filaments in water. The bright region, as indicated by *arrows*, is 50 nm. The *arrows* at the *top right-hand side* are defined as the directions of wavevector **k** and electric field vector **E**

optical imaging in water was also done by c-mode SNOM under NFO intensity feedback control as in air. The filament sample was fixed on a glass substrate and was fitted with an acrylic ring. The columnar bath with an inner diameter of 10 mm and a height of 2 mm was filled with water during the observation. The sample–probe distance was estimated to be less than 30 nm. The bright segments correspond to the filaments. The FWHM of the

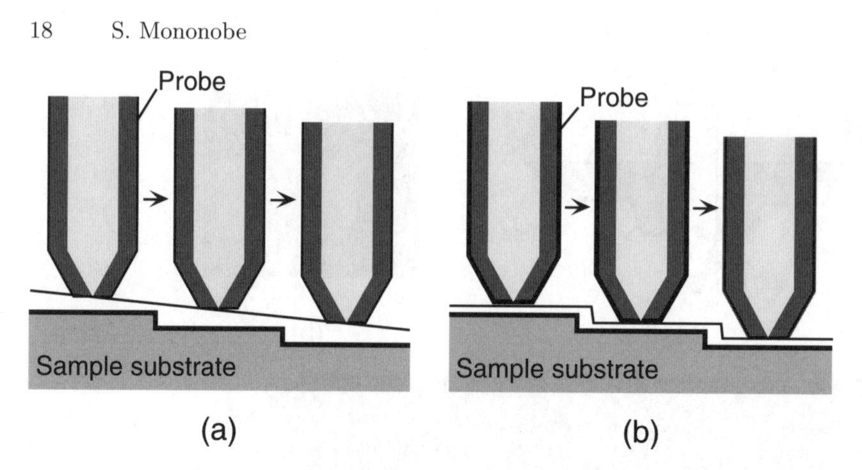

Fig. 16. (**a**) constant-height mode and (**b**) constant-distance mode

bright region as indicated by arrows is 50 nm. The c-mode SNOM employing the probe with a protruding tip is effective in the near-field imaging of biological sample in water.

Figures 15c and d were obtained with c-mode SNOM operated in the constant-height mode as schematically shown in Fig. 16a. The monotonous variation as shown in Fig. 14 was used to select the initial distance. To compensate the tilt effects of the substrate, the time constant of the feedback loop was carefully adjusted so that the probe followed only sample features larger than 1 μm. When the c-mode SNOM was operated in the constant-distance mode (Fig. 16b), such highly resolved images could not be obtained.

3.4 c-mode SNOM Images of Microtubules

Zvyagin et al. [31] isolated microtubules from the brain of a pig, and observed with c-mode SNOM using a protrusion-type probe with a foot diameter of 30 nm. For this observation, a c-mode SNOM with s-polarized incident light was used. To regulate the sample–probe distance, a shear-force feedback system [1, 2] was added to c-mode SNOM. In the shear-force feedback system, the differential force and frictional force are detected by dithering the probe at its resonance frequency at some amplitude, and the amplitude of dithering is measured by an optical interference technique. Figure 17 shows the variation of the amplitude of dithering of the protrusion-type probe as a function of the sample–probe distance. The inset is the resonance frequency spectrum.

Figure 18a shows the transmission electron micrograph of microtubules. The dark region seen in the middle of a strand corresponds to microtubules. The width of the microtubules, as indicated by arrows, is 25 nm. Figure 18b shows a shear-force topographic image of microtubules. The sample–probe distance was kept at less than 5 nm during the scanning. Here, the long bright structure and the round ones correspond to a microtubule and protein aggregates, respectively. From this high-resolution imaging result, the

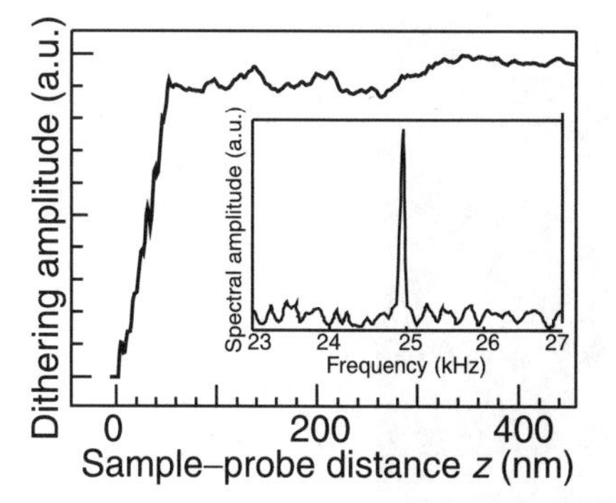

Fig. 17. Variation of the dithering amplitude of the probe as a function of the sample–probe distance. The *inset* is the resonance frequency spectrum of the dithering probe

protrusion-type probe is very effective for high-resolution shear-force microscopy. Figure 18c shows the near-field optical image of the rectangular region in Fig. 18b, which is obtained in the constant height mode without using shear-force feedback. The sample–probe distance was about 30 nm.

3.5 Near-Field Spectroscopic Investigation of Semiconductor Quantum Dots Under Extremely Low Temperature

Toda et al. [32] carried out a spectroscopic study of GaAs quantum dots by low-temperature SNOM employing a protrusion-type probe. Figures 19a shows schematic illustration of the microscope, the probe, and the GaAs quantum dots grown on SiO_2-patterned GaAs substrates with metalorganic chemical vapor deposition. Here, the probe has a foot diameter of 100 nm and a cone angle of 40°. The length of the protruding tip is about 130 nm. The quantum-dot pattern has a size of 190 nm $\times$160 nm $\times$12 nm and a separation of 2 μm. The excitation light is delivered via the fiber probe with a length greater than 2 m from the outside of the cryostat to the sample. The photoluminescence spectrum as shown in Fig. 20a was obtained by positioning the protruding tip 200 nm above the top of the quantum-dot at a low temperature of 18 K. This corresponds to the spectrum from the carriers excited in the whole quantum dots structure including the regions A–C. Figures 20b and c show photoluminescence spectrums with tip position less than 10 nm above the region of the quantum dots and that of SiO_2 mask, respectively.

Figures 20d–f show monochromatic photoluminescence images obtained in the constant distance-mode at 18 K with the energy regions indicated by the arrows labeled A–C, which originated from the three regions of the GaAs bulk, GaAs quantum dots, and GaAs quantum well, respectively. Figure 20g is a shear-force topographic image obtained simultaneously. The protrusion-type fiber probe with the long protruding length is convenient for scanning such bumpy sample surfaces.

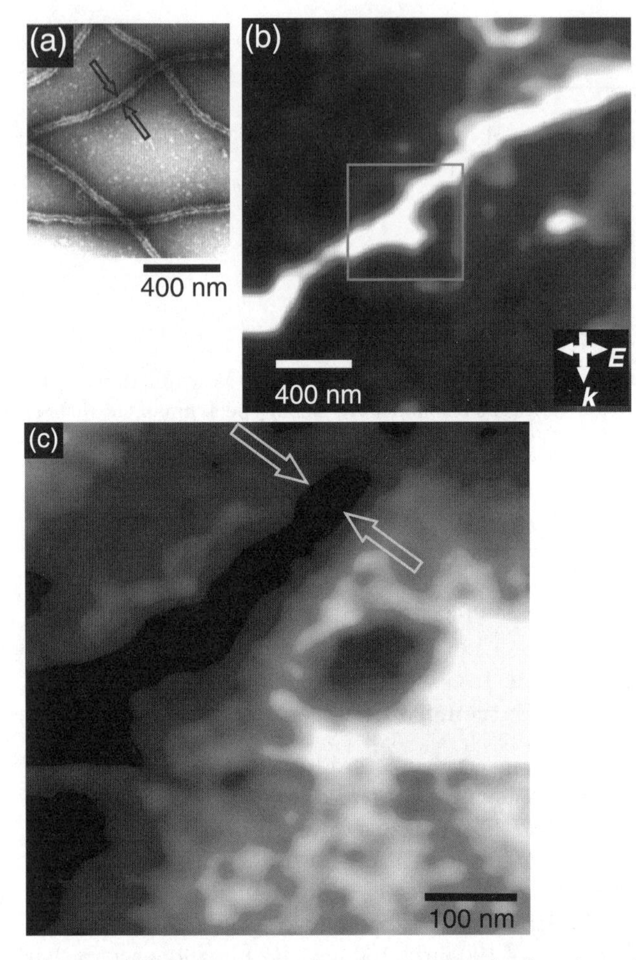

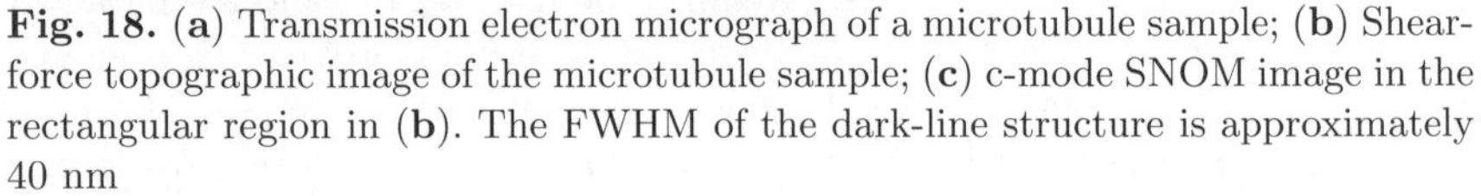

Fig. 18. (**a**) Transmission electron micrograph of a microtubule sample; (**b**) Shear-force topographic image of the microtubule sample; (**c**) c-mode SNOM image in the rectangular region in (**b**). The FWHM of the dark-line structure is approximately 40 nm

For this near-field imaging, other probe types were also used, which we fabricated by pulling and etching the highly GeO_2-doped fiber. In the heating-and-pulling process, the fiber was tapered with the highest pulling strength by a commercial puller (Sutter, P-2000) so as to have a dip structure at the taper end. By the etching in a BHF with a volume ratio of 10:1:1, the dip was flattened, and then, the core is protruded from the pure silica clad. The tapered fibers were aluminized by the vacuum evaporation method. Some of the probes had throughputs higher than the protrusion-type probes. However, we could not fabricate high-throughput probes with reproducibility because it is difficult to control the sizes of the dip generated at the pulling. Figures 20a

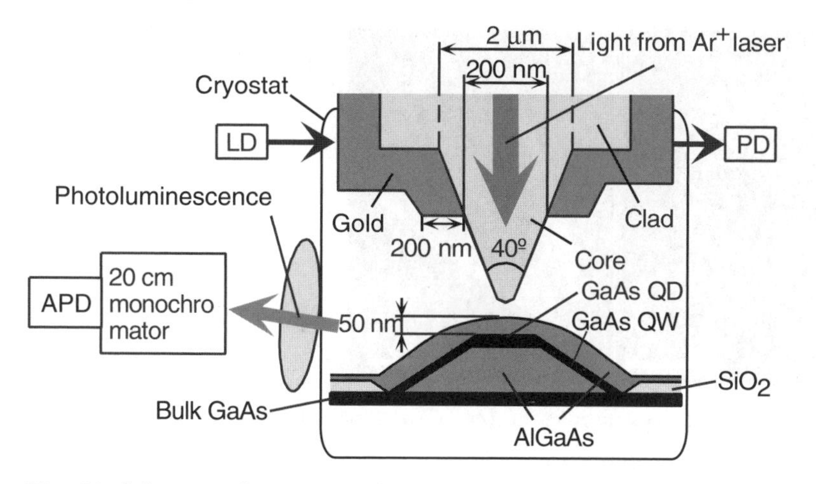

Fig. 19. Schematic illustrations of i-mode SNOM developed for photoluminescence spectroscopy of GaAs quantum dots at a low temperature of 18 K

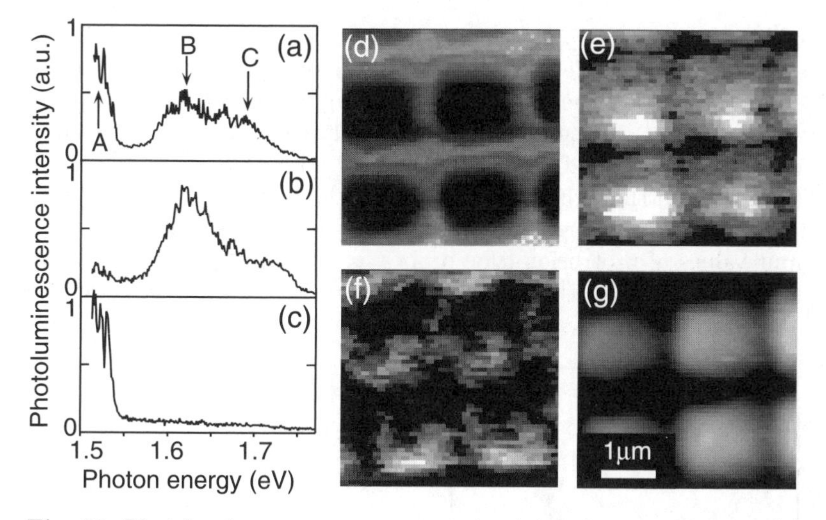

Fig. 20. Photoluminescence spectra on maintaining the protruding tip (**a**) 200 nm above the quantum dots, (**b**) less than 10 nm above the quantum dots, (**d**) less than 10 nm above the SiO_2 mask; (**c–f**) Monochromatic photoluminescence images at the energy regions labeled A–C, respectively, and simultaneously obtained topographic image

and b show scanning electron micrographs of a pulled-and-etched probe and its magnified view. The dotted lines represent the cross section of the GeO_2-doped core.

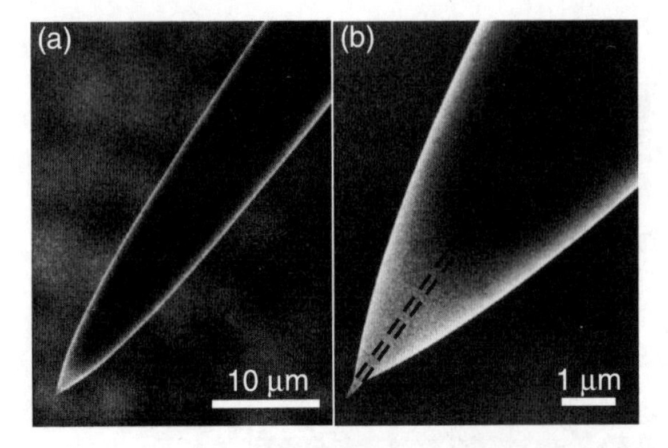

Fig. 21. Scanning electron micrographs of (a) a pulled and etched probe and (d) its magnified view

3.6 Transmission Efficiencies of the Protrusion-type Probes

In near-field spectroscopic applications, one has to cope with near-field signals that are too small. In i-mode SNOM, the transmission efficiency of the probe is defined as the ratio of the output power to input power coupled into the fiber probe. The low efficiency is caused by a large guiding loss, which is generated along the tapered core due to optical interaction with the metal clad. Therefore, the throughput is affected by the length of the metallized tapered core or the cone angle. Figure 22 shows variations of the measured throughput values of protrusion-type probes for the different cone angles of 20° and 50° as a function of the foot diameter of the protrusion. Here, the

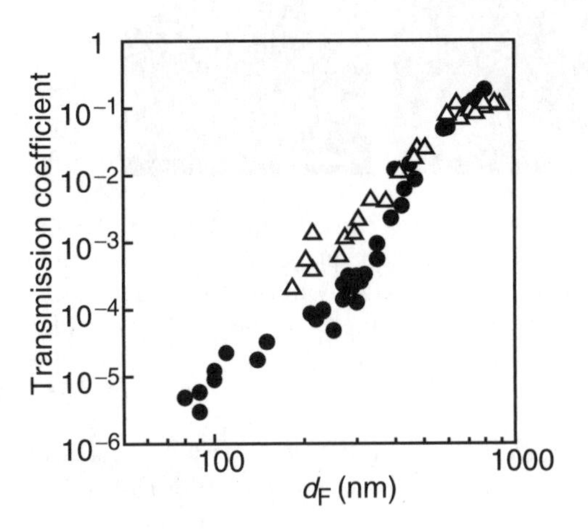

Fig. 22. Dependencies of the transmission efficiency of two protrusion-type probes with a cone angle of $\theta = 20°$ (*closed circles*) and $\theta = 50°$ (*open triangles*) on the foot diameter

closed circles and open triangles correspond to the variations for the cone angles of 20° and 50°, respectively. To collect the scattered light from the apex of the probe, an objective lens with a numerical aperture of 0.4 was inserted in front of the photodetector. At $d_F = 200$ nm, The throughput of the probe with a large cone angle of 50° is 10 times larger than that with $\theta = 20°$. Increasing the cone angle enhances the throughput. For a near-field spectroscopic application requiring high throughput, the tapered probe with a large cone angle, i.e., 50–90° should be fabricated by a selective-etching method.

4 Metal-Dielectric-Metal-Coated Fiber Probe and Near-Field Imaging of DNA Molecules

For imaging application of single biomolecules, i.e., a single DNA string with a lateral width of 4 nm and a height of 2 nm, one should fabricate a near-field probe with high-resolution capability beyond 10 nm. Recently, we fabricated a specially designed probe with a metal-dielectric-metal coating for nanometer-level-resolving SNOM, and succeeded in obtaining a c-mode SNOM image of deoxyribonucleic acid (DNA) molecules [33].

4.1 Ag-MgF_2-Al-Coated Fiber Probe

Figure 23a shows a schematic image of a fabricated probe with metal-dielectric-metal coating. The probe consists of two tapered structures with the initial taper with a cone angle of 62° and a tapered apex region with

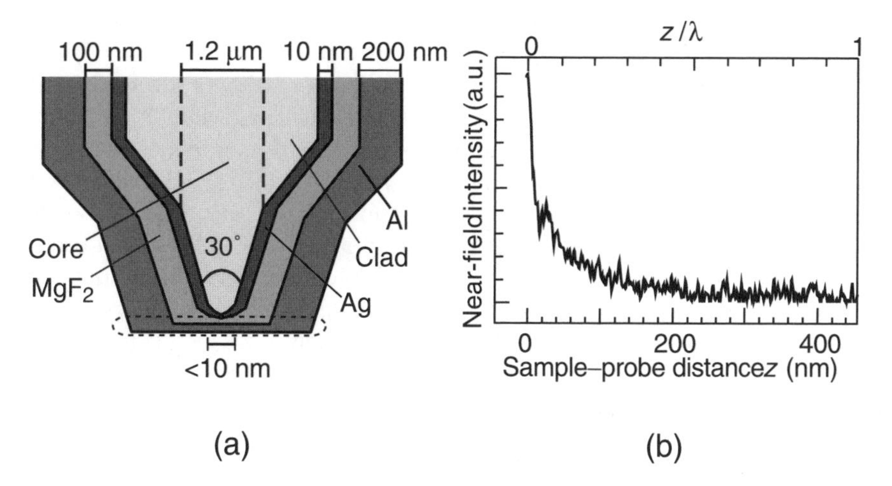

Fig. 23. (**a**) Ag-MgF_2-Al-coated fiber probe; (**b**) Variation of the optical near-field intensity as a function of the sample–probe distance

a cone angle of 30°. We call this tapered fiber a pencil-shaped probe. The method for fabricating the pencil-shaped probe with a metal-dielectric-metal coating involves two steps: first sharpening of an optical fiber a into pencil-shaped structure and then coating in a vacuum evaporation unit.

Fabrication of a Pencil-shaped Probe

The pencil-shaped probe with a cone angle of 30° and an apex diameter less than 10 nm was fabricated with almost 100% reproducibility based on selective etching of a multistep index fiber with the GeO_2-doped silica core, pure silica clad, and a F-doped silica support. This fiber is specially designed and fabricated by the VAD method to produce pencil-shaped probes and triple-tapered probe. For details of the etching process, refer to the Appendix.

Ag-MgF_2-Al Coating

The pencil-shaped probe is coated in a vacuum evaporation unit. The coating consists of three layers with an inner thin coating of silver (Ag), a middle layer of magnesium fluoride (MgF_2), and an outer aluminum coat. These materials were chosen due to the following facts:

- The inner thin layer of Ag is mainly for enhancing the scattering efficiency of the three-dimensionally localized near field around the sample. Ag has a high reflectivity at the region of the wavelength used, and has a higher dielectric constant than silica. Furthermore, based on calculated results of Novotony et al. [42], the field confinement for a tapered probe with a thin silver coating is highly concentrated in both the lateral and vertical or z directions. The field confinement of the Ag-MgF_2-Al-coated probe in the z direction could be inferred directly from Fig. 23b, which indicates strong near-field intensity enhancement very close to the sample surface.
- The middle layer of MgF_2 has a lower refractive index of 1.3 compared with 1.5 of the sharpened core and forms an efficient clad. It not only prevents the deterioration of the inner layer of silver on exposure to air but also plays some role in suppressing the propagating light. Based on experimental experience, adding MgF_2 improves the scattering efficiency of the probe and also the robustness of the probe due to its soft nature.
- The outer Al coating with a fairly large thickness of 200 nm in comparison with the skin depth, which is a few tens of nanometers, is coated in order to suppress background unwanted scattered light.

4.2 Sample Preparation

The double-stranded plasmid DNA with a ring structure (pUC18 and 2868 base pairs) was used as the sample. It was diluted with distilled water to a final concentration of 5 ng/μL. A quantity of 2 μL of this was dropped onto the center of a wet-treated sapphire substrate with an ultrasmooth surface. Next, the wet-treated surface was blown dry with compressed air.

The sapphire substrate was produced by an epitaxial growth method [43], which is stable both in air and in water for a considerable duration. The surface has a staircase-like step structure whose step height and terrace width can be carefully controlled. As the sapphire surface itself is hydrophobic, it is necessary to make it hydrophilic for a stronger adsorption of organic DNA to the surface. For this purpose, the sapphire surface was treated with a sodium diphosphate solution with a concentration of 0.5 mol/L for a few minutes, and then rinsed with distilled water. After being dried in air, the DNA solution was dropped onto the substrate.

4.3 DNA Images

Figure 24a shows the image of DNA on the ultrasmooth sapphire surface obtained using a noncontact-mode atomic force microscope [43]. The bright-ring structures correspond to DNA molecules. The observed width of a single strand, as indicated by open arrows, is estimated to be around 10 nm.

The experimental system of c-mode as appears in Fig. 13 was used for near-field optical imaging of single DNA molecules. The DNA sample is mounted on the prism, and the optical near field is sensitively picked up by the Ag-MgF_2-Al-coated probe. Figure 24b shows the c-mode SNOM image, obtained under near-field optical intensity feedback control. The pixel size is 5 nm $\times$ 5 nm. The arrows at the top left-hand side of the figure indicate the propagation $\mathbf{k}$ and electric field $\mathbf{E}$ vectors of the incident light with a 680 nm wavelength. The coiled loop and single strands of DNA are observed in this figure. The FWHM of the intensity variation at the portion, as indicated by the line, is around 20 nm. Figure 24c shows the magnified view of the single strands of DNA. The pixel size is 2 nm $\times$ 2 nm. The FWHM, as indicated by the arrows, is about 4 nm. This nanometric spatial resolution reflects the special care taken during preparation of the sample and of the probe to avoid unwanted scattered light.

We tried to obtain near-field optical images of DNA molecules by employing the protrusion-type probe with a foot diameter of 30 nm. However, successful results could not be obtained. The metal-dielectric-metal-coated probe tip with a layer of thin silver coating acts as a scattering probe much more sensitive than a dielectric tip protruding from metal.

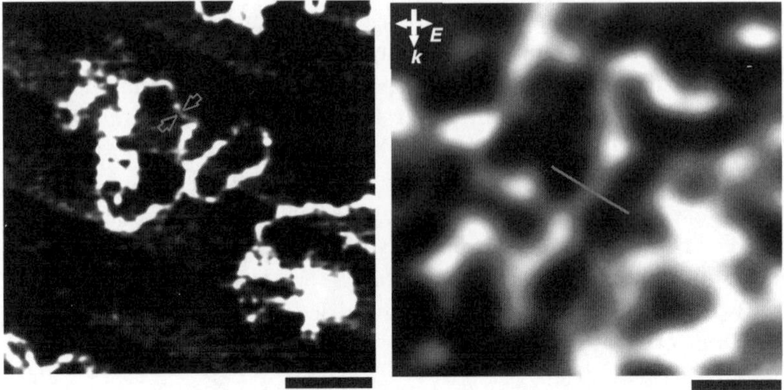

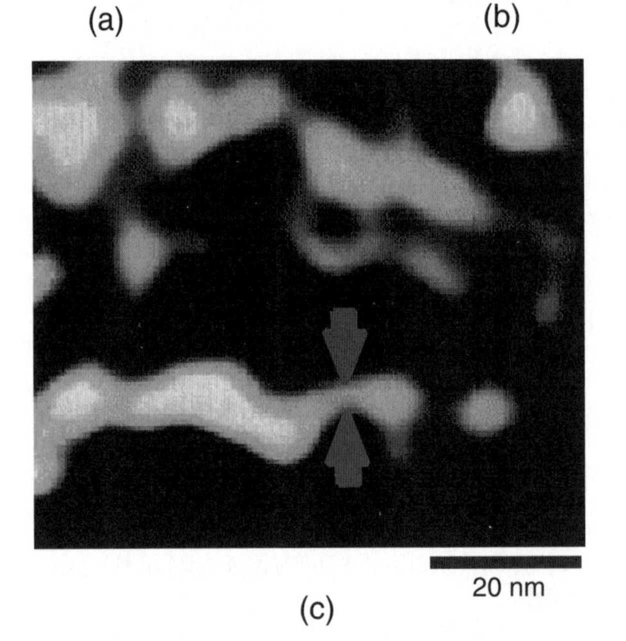

Fig. 24. (**a**) AFM image and (**b, c**) near-field optical images of a DNA sample, obtained with c-mode SNOM employing the fiber probe in Fig. 23a

5 Apertured Probes for Near-Field Imaging of Dye-Doped Samples

In investigating biological samples with i-mode SNOM, it is an effective method of enhancing the optical contrast to attach the sample with absorbing and fluorescing dye labels. For high-resolution imaging of a dye-labeled sample, one has to fabricate the apertured probe with an aperture diameter of a few tens of nanometers. Maheswari et al. produced a fiber probe with an

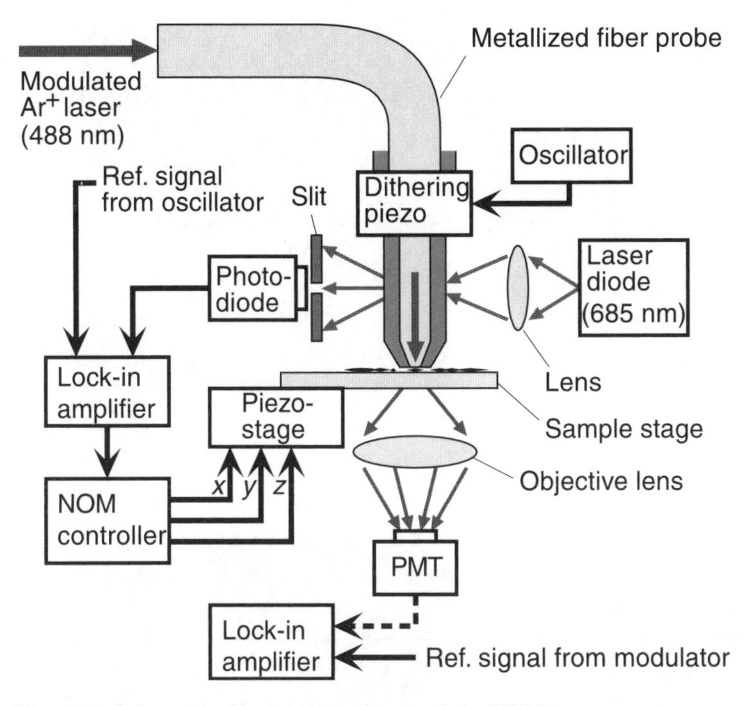

Fig. 25. Schematic illustration of an i-mode SNOM

aperture diameter of around 15 nm and obtained near-field images of dye-labeled neurons of rats with an i-mode system under the shear-force feedback control. Figure 25 shows the schematic illustration of the used i-mode system. Here, the light from an argon ion laser with a 488 nm wavelength is coupled into the fiber probe to generate a localized optical near field around the apex region of the probe. The sample is mounted on the xyz piezo-stage. On approaching the sample surface very close to the apex of the probe, the optical near field generated around the probe is scattered and detected by the sample fine structures. The scattered light is detected by a photomultiplier tube through collection optics such as the objective lens with a high numerical aperture. To regulate the sample–probe distance, a shear-force feedback system is added consisting of the 685-nm laser diode, dithering probe, and photodiode.

Figure 26a shows the schematic illustration of the fabricated probe. The cone angle is 20°. To fabricate the probe, a dispersion-compensating fiber was consecutively etched in three BHFs with volume ratios of 1.7:1:1, 10:1:1, and 10:1:120, for 60, 75, 2 min, respectively, and then was metallized with 5-nm thick chromium and gold with a thickness of 120 nm by a vacuum evaporation unit.

Figure 26b shows the i-mode near-field image of a neuron labeled with toluidine blue. Here, the sample–probe distance is less than 50 nm. The wide

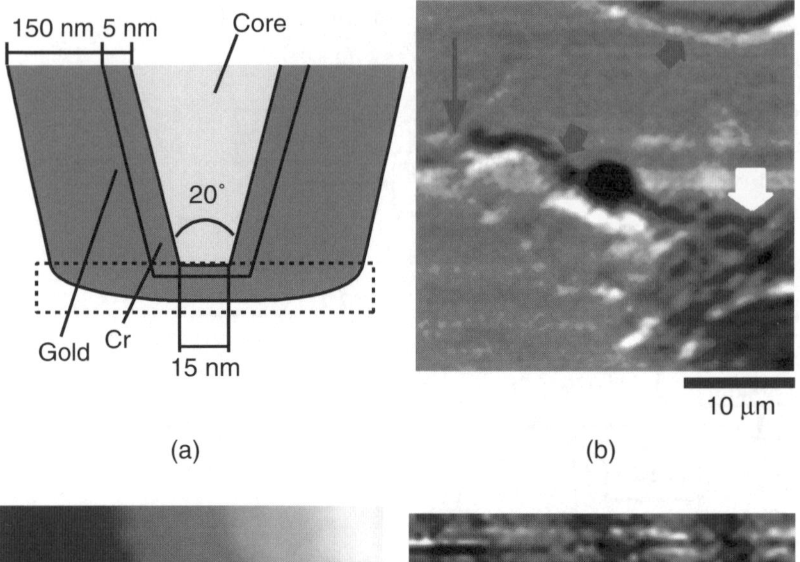

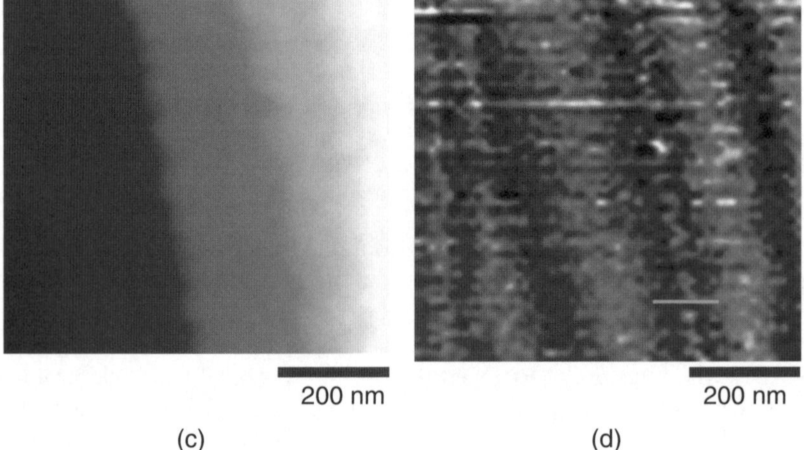

Fig. 26. (**a**) Cross-sectional illustration of the fabricated metallized fiber by the vacuum evaporation method. Its top region enclosed by the dotted rectangular was removed by rubbing with the substrate under shear-force feedback control before the near-field imaging; (**b**) near-field optical image of neurons of rats. The magnified view of (**c**) shear-force image and (**d**) i-mode SNOM images of the same area

dark arrows and the bright arrow correspond to the neural process and the cell body, respectively. Before this imaging, the metal covering the top region of the probe, enclosed by the dotted rectangular in Fig. 26a, was removed by rubbing with the sample substrate under shear-force feedback control. Figures 26c and 26d correspond to the simultaneous shear-force and near-field images of the magnified region, as shown by the long dark arrow in the neuron process in Fig. 26b. The fringe-like structures are seen in the near-field image of Fig. 26d, however, they do not appear in the topographic image. This shows that the fringe-like structures labeled with toluidine blue lie just underneath the cell membrane on the surface. The structures are identified as microtubules, which are the main constituent elements of the neural process. The FWHM of near-field intensity variation at the portion, as indicated by the line in Fig. 26d, is estimated to be 26 nm, which agrees with the typical diameter of 25 nm of a single microtubule.

Recently, another apertured probe with a large cone angle, i.e., 50–90° has been chosen for performing i-mode and i-c mode imaging applications including single-dye-molecule detection because the probe has a much higher throughput than the tapered probe as shown in Fig. 26a. Hosaka et al. [35] fabricated a double-tapered probe [44] with a cone angle of 90° and an aperture diameter of a few tens of nanometers as shown in Fig. 27a, and succeeded in obtaining near-field fluorescence imaging of dye molecules with a high resolution of around 10 nm. Figure 27b shows the near-field fluorescence image of single Rhodamin 6G molecules dispersed on the quartz glass at 700 ± 20 nm. In this experiment, the dye was excited by a He-Ne laser with a 633 nm wavelength. The bright spots correspond to emitting single dye molecules. The estimated resolution is around 10 nm. For details of the double-tapered probe see Sect. 6.

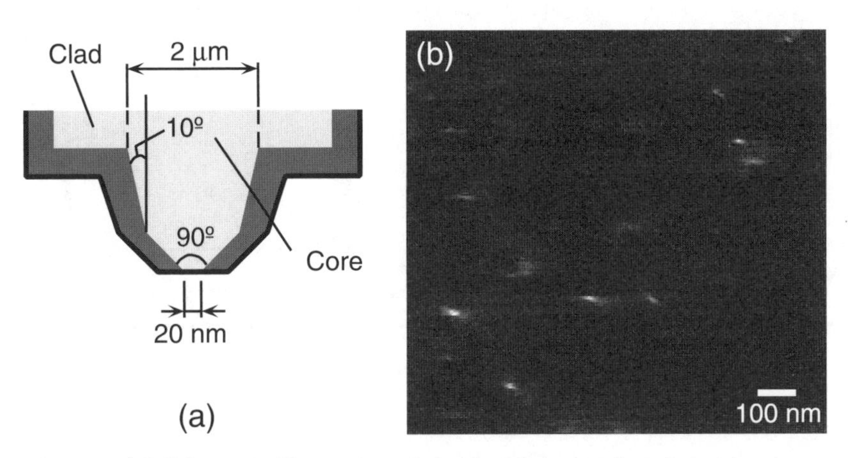

Fig. 27. (**a**) Schematic illustration of the double-tapered probe with a cone angle of 90°; (**b**) i-c mode near-field fluorescence image of single Rhodamin 6G molecules dispersed on a quartz substrate at 700 ± 20 nm

6 Double-Tapered Fiber Probe and Spectroscopic Applications

SNOM is a powerful tool for studying semiconductor devices with submicrometer and subnanometer sized structures by imaging and spectroscopy. For photoluminescence investigation of a semiconductor device, i-c mode SNOM whose resolution is not affected by carrier diffusion effect should be chosen. In i-c mode SNOM (Fig. 1c), the sample is effectively excited by the optical near field generated around the probe in the i-mode fashion, and the near-field photoluminescence of the sample is scattered by the probe in the c-mode fashion. To perform imaging and spectroscopic applications by illumination-collection hybrid mode SNOM, high throughput probes must be fabricated. Saiki et al. [44] have developed a double-tapered probe with high throughput, and have performed a spectroscopic study of semiconductor devices by i-mode SNOM and i-c mode SNOM employing the probes. In this section, double-tapered probe and spectroscopic applications are described.

6.1 Double-Tapered Probe

Figures 28a and b show the cross-sectional profiles of the protrusion-type double-tapered probe and apertured-type double-tapered probe. Here, θ_1 and θ_2 ($< \theta_1$) are the cone angles of the first and second taper, respectively. λ is the cross-sectional diameter, which agrees with the optical wavelength in the core. d_B is the base diameter of the first taper, which is larger than

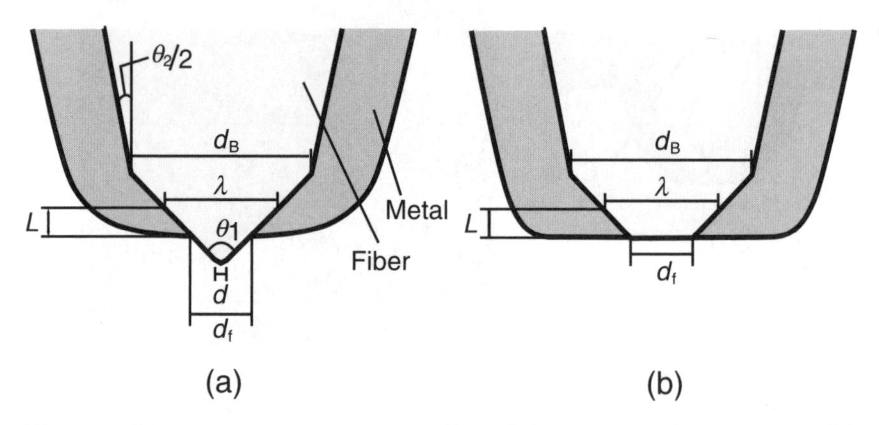

Fig. 28. Schematic cross-sectional profiles of double-tapered probes with (**a**) a protruding tip from metal and with (**b**) an aperture. θ_1, θ_2, cone angle of the first and second tapers, respectively; d, apex diameter; d_f, foot diameter in (**a**) and aperture diameter in (**b**); d_B, base diameter of the first taper; λ, optical wavelength in the core; L, length of the lossy part between the cross-sectional diameters of λ and d_f

the wavelength size λ, and d and d_f are the apex and foot diameters of the protrusion, respectively.

As discussed in Sect. 3.6, the first cone angle of the double-tapered probe has to be increased to enhance the throughput of the probe. Since the light entering the tapered core is strongly attenuated by the metal film in the subwavelength cross-sectional portion, by making the base diameter the value of λ or more, the double-tapered probe with the first cone angle of 90° has a transmission coefficient as high as the single-tapered probe with a cone angle of 90°. Based on FDTD simulation [45], the double-tapered probe can have a collection efficiency higher than the single-tapered probe. The double-tapered structure is suitable for coupling the scattered light into the core.

Forming a Double-Tapered Tip

The method for fabricating the probe involves two steps: (A) controlling the clad diameter and forming the first taper; (B) forming the second taper, which are based on selective etching as described in Sect. 2.3. Figure 29a shows the method schematically. In step A, by immersing the fiber in BHF with X_A:1:1 (where $X_A > 1.7$), the core is protruded from the clad with the cone angle θ_1 represented by

$$\sin\frac{\theta_1}{2} = \frac{R_{1A}}{R_{2A}} . \tag{8}$$

In step B, immersing the fiber in BHF with 10:1:1 the core is tapered with cone angles of θ_1 and θ_2, where

$$\sin\frac{\theta_2}{2} = \frac{R_{1B}}{R_{2B}} . \tag{9}$$

The base diameter d_B is represented by

$$d_B = 2r_1(1 - \frac{T_B}{\tau}) \qquad (T_B < \tau) , \tag{10}$$

which is controlled by varying the etching time T_C. Here, τ is defined by (4). The clad diameter D is given by (7).

In actual conditions as shown in Table 4, a probe with a double-tapered core with cone angles of $\theta_1 = 90°$ and $\theta_2 = 20°$, and a base diameter of $d_B = 500$ nm was obtained. Figures 29b and c show SEM micrographs of the tapered core region of a double-tapered probe and its magnified top region.

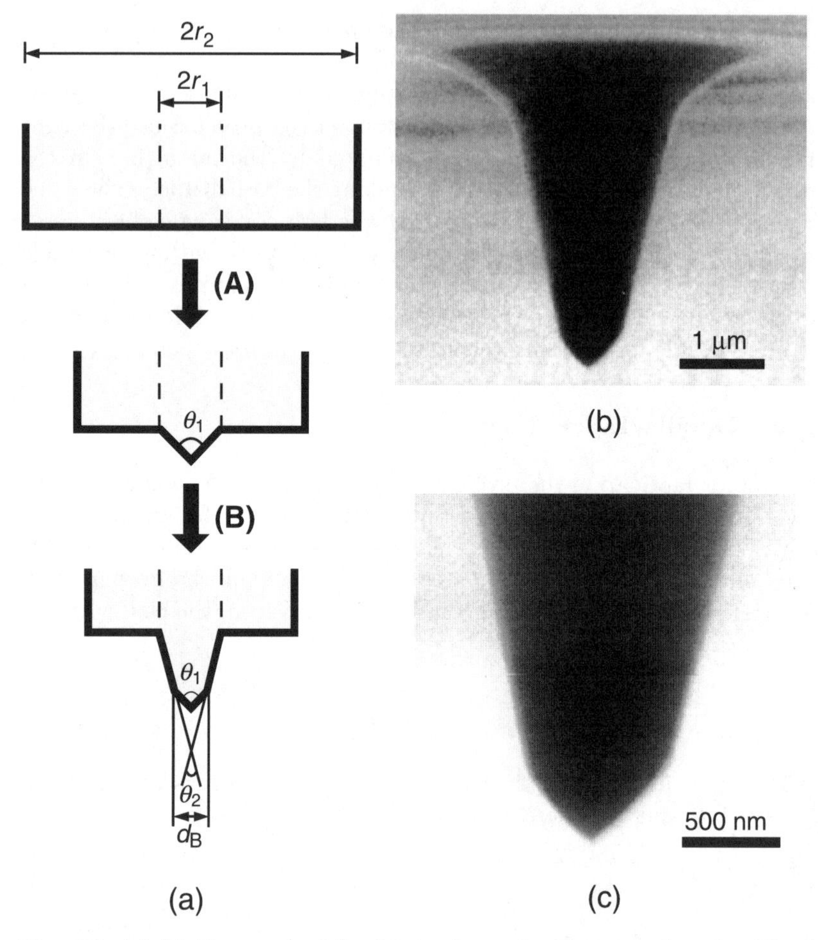

Fig. 29. (**a**) Etching method for fabricating a double-tapered probe. The fiber is consecutively etched in buffered HF with volume ratios of X_A:1:1 ($1.7 < X_A < 10$) and 10:1:1 in steps A and B, respectively. Here, $2r_1$ and $2r_2$ are the core and clad diameters, respectively. θ_1 and θ_2 are the first and second cone angles of the double-tapered probe; SEM micrographs of (**b**) a double-tapered core and (**c**) the magnified top region. $\theta_1 = 90°$; $\theta_2 = 20°$; $d_B = 500$ nm

Table 4. Conditions for fabricating a probe as in Fig. 29

Step	Volume ratio of BHF	Etching time (min)
A	1.8:1:1	70
B	10:1:1	40

6.2 Near-Field Photoluminescence Image of Lateral p-n Junctions Obtained with the i-c Mode SNOM

With SNOM employing a double-tapered probe, Saiki et al. [36] investigated a GaAs-based lateral p-n junction, and obtained its near-field photoluminescence image with i-c mode SNOM. Figures 30a and b show the schematic illustrations of the double-tapered probe and experimental setup of the i-c mode SNOM, respectively. Here, the probe was produced based on selective resin coating. The foot diameter and cone angle of the double-tapered probe is 200 nm and 50°, respectively. The GaAs sample is fixed on the inclined substrate to avoid the bumpy sample surface with the edge of the metallized clad on scanning. The sample is excited by laser light with 633 nm through the double-tapered probe. The laser power coupled into the fiber is 0.5 mW. The sample–probe distance was controlled by the shear-force feedback technique to scan the sample. The sample is fabricated by growing a silicon-doped GaAs layer on a semi-insulating GaAs (111)A substrate with (311)A slopes [46]. The thickness of the silicon-doped layer is around 1 μm.

Figures 31a and b show the shear-force topographic image and near-field photoluminescence image, which were simultaneously obtained. Because photoexcited electrons and holes are separated and drifted by the internal electric field in the upper and lower junctions, the emission intensity is lower than those of the other regions as seen in Fig. 31b. Furthermore, in the lower junction region, some bright portions where the photoluminescence intensity increased locally is observed. From the intensity variation in this region, the spatial resolution is estimated to be 200 nm or less. Recently, near-field photoluminescence images of single GaAs dots and a GaNAs layer

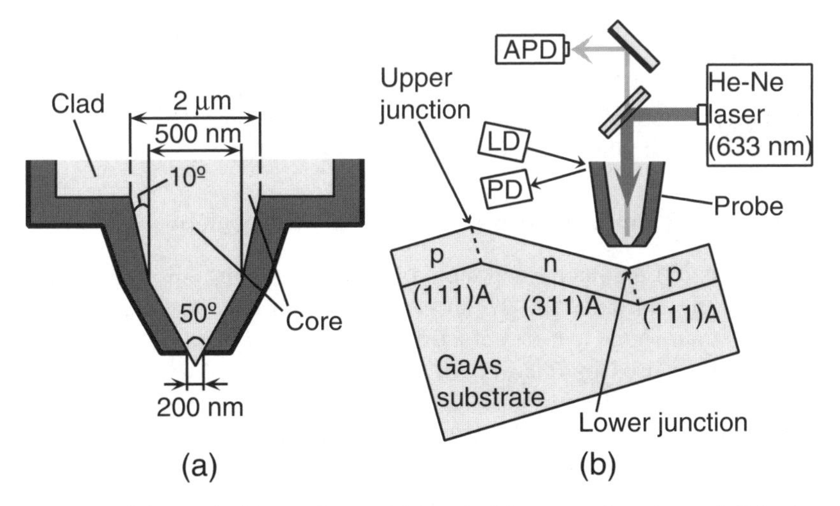

Fig. 30. Schematic illustrations of (**a**) a double-tapered probe and (**b**) a i-c mode SNOM for investigating the lateral p-n junction

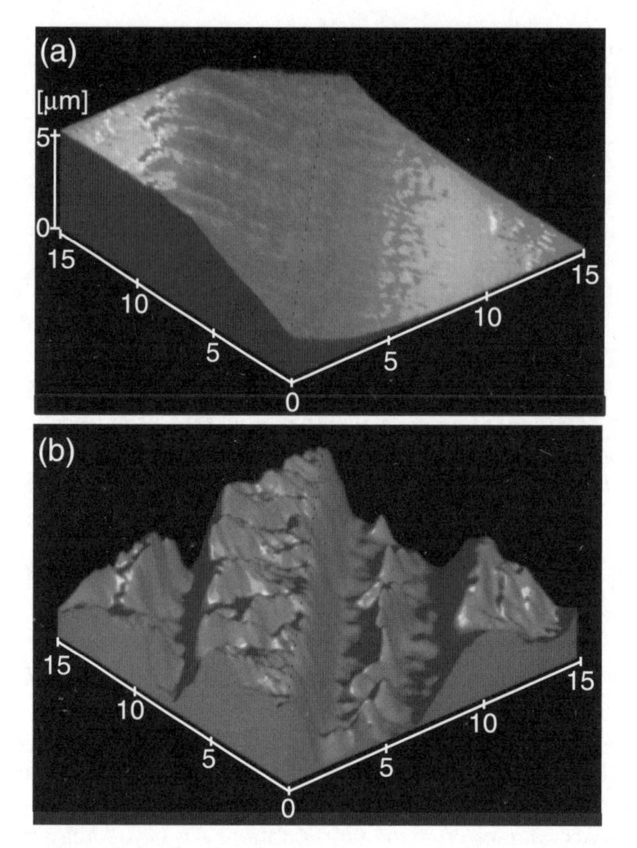

Fig. 31. (a) Shear-force topographic and (b) i-c mode SNOM images of the GaAs lateral p-n junction

have been obtained with i-c mode SNOM employing a double-tapered probe with an aperture diameter of 20–30 nm, and a resolution of 20–30 nm has been achieved. For details of these studies, see Saiki [47].

6.3 Near-Field Raman Spectroscopy of Polydiacetylene

With i-mode SNOM employing a double-tapered probe with an aperture diameter of 100 nm, Narita et al. [37] succeeded in obtaining two-dimensional mapping of Raman-signal features of a tabular polydiacetylene single crystal. The sample was excited by 532-nm Nd:YAG laser with a power of 10 nW. Near-field Raman spectra were measured at 10×10 points in a portion with an area of $1\,\mu m \times 1\,\mu m$. Figure 32a shows a near-field Raman spectra of polydiacetylene. Here, two Raman peaks L and S are observed at 1457 cm^{-1} and 1520 cm^{-1}, which relate to the C=C bond and reflect the difference in the number of successive *cis*-bonds. Peaks L and S originate from a longer successive bond and shorter ones, respectively. Figure 32b shows a two-dimensional

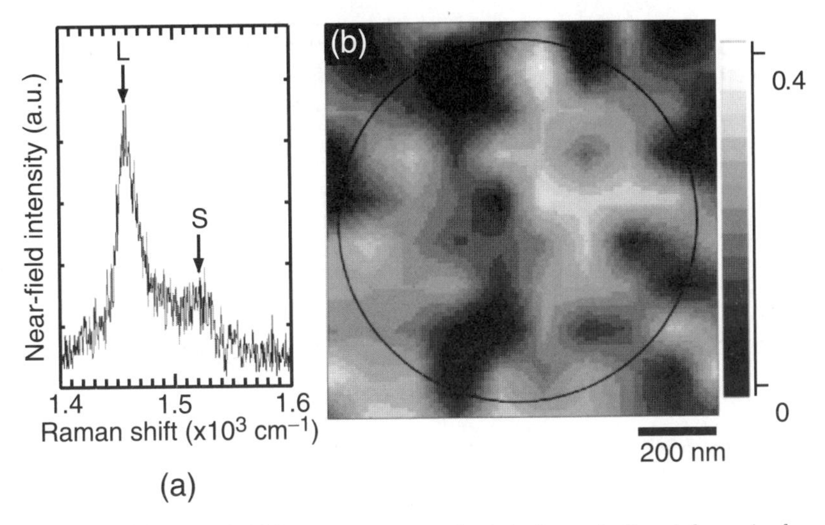

Fig. 32. (**a**) Near-field Raman spectra of a tabular polydiacetylene single crystal; (**b**) Two-dimensional mapping the relative near-field intensity at peak S to peak L

mapping of the relative Raman intensity of peak S at 1520 nm to peak L at 1457 nm.

7 Pure Silica Core Fiber Probes and Ultraviolet Applications

Using GeO_2-doped fibers such as the dispersion-compensating fiber and the double-clad fiber, one can fabricate near-field probes as appeared in Figs. 15a, 19a, 21a, 23a, 26a, 27a, and 30a and carry out SNOM applications in the visible and infrared regions. However, it is difficult to employ such probes for ultraviolet (UV) near-field optical microscopy applications because the GeO_2-doped silica fiber has strong guiding loss based on Rayleigh scattering and moreover, has UV absorption, luminescence, and Raman scattering that originate from defects in the GeO_2. In order to overcome this difficulty and to realize UV-SNOM, we have proposed fiber probes with the pure silica core. Figure 33 shows optical transmission spectra of a GeO_2-doped fiber with an index difference of 1% and a fiber with the pure silica core and the fluorine-doped clad. It is found that the pure silica fiber probe has high transmittance in a wide region from near UV to near infrared. To realize pure silica fiber probes, we have developed methods to fabricate the following pure silica core probes:

- Single-tapered probe with a large cone angle of 120°
- Pulled and etched probe with the tapered core and clad [25]
- UV triple-tapered fiber probe [48]

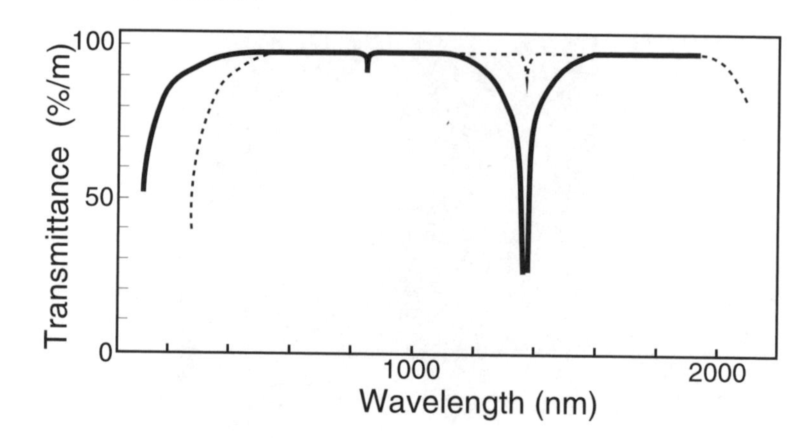

Fig. 33. Optical transmission spectrums of a GeO_2-doped core fiber with a refractive index difference of 1% (*dotted curve*) and a pure silica core fiber (*solid curve*)

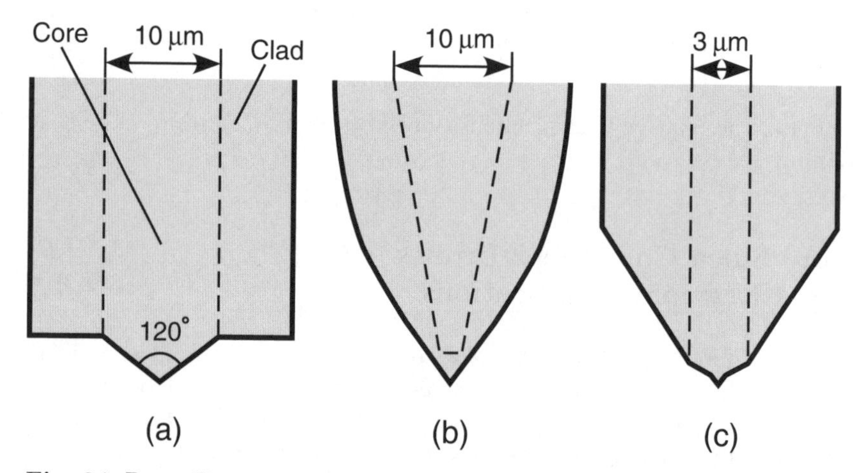

Fig. 34. Pure silica core probes with (**a**) the conical core protruded from the flat clad end (**b**) tapered clad and core, and (**c**) triple-taper structure

as schematically shown in Fig. 34a–c, respectively. The single-tapered fiber is fabricated by immersing a 1300-nm pure silica fiber (PSF) (Sumitomo, PS1) with a core diameter of 10 µm in a BHF as described in Sect. 2.3. Figure 35 shows a scanning electron micrograph of the single-tapered fiber with the pure silica core and a cone angle of 120°. We have tried to develop UV single-mode fibers with a cutoff wavelength of less than 400 nm in order to fabricate UV single-tapered fiber. Details of these fibers will be discussed elsewhere. Using the pulled and etched probe and triple-tapered probe, we have obtained UV near-field photoluminescence images of polysilane and a near-field Raman spectrum of a silicon crystal. In the following, triple-tapered and pulled and etched probes are described.

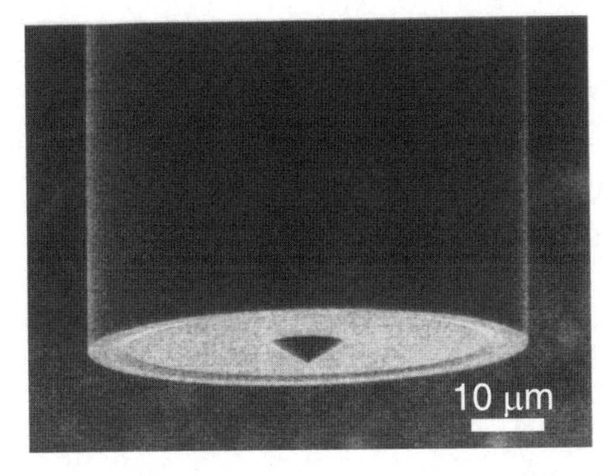

Fig. 35. Scanning electron micrograph of a tapered fiber with a pure silica core, obtained by immersing a 1300-nm fiber with a core diameter of 10 μm for 240 min in a BHF with a volume ratio of 10:1:1. The cone angle is 120°

7.1 UV Triple-Tapered Probe

To fabricate a UV triple-tapered fiber probe, we recently developed a UV multistep index fiber [48] with the double core involving a subwavelength core and a pure silica core. In the following text, this fiber is called a double-core fiber. For triple-tapered probes with the GeO_2-doped core, refer to the Appendix.

Fabrication of UV Triple-Tapered Probe Based on Selective Etching of a Double-Core Fiber

The inset in Fig. 36 shows the index-difference profile of the double-core fiber. The diameters of sections 1–5 are 100 nm, 2.9 μm, 40 μm, 42 μm, and 125 μm, respectively. Sections 3–5 correspond to the clad region. Sections 3 and 5 are made of low fluorine-doped silica with an index difference of −0.35%. Section 4 is made of the high fluorine-doped silica with an index difference of −0.94%. The core region consists of sections 1 and 2, which are made of GeO_2-doped silica and pure silica, respectively. The index difference of section 1 to section 2 is estimated to be around 0.7%. The diameter of section 1 is only 100 nm. Since 99.9% of the cross-sectional area of the core region is occupied by the pure silica, the double-core fiber has transmittance as high as that of the pure silica fiber as shown in Fig. 33.

To fabricate this double-core fiber, we fabricated a preform glass rod by a combination of vapor-phase axial deposition and plasma-activated chemical vapor deposition and then drew the 125-μm diameter fiber by heating the preform. Table 5 summarizes the parameters for drawing the fiber.

As shown in Fig. 36, the double-core fiber is tapered by a selective etching method, which involves three steps: (A) hollowing of sections 1 and 4, (B)

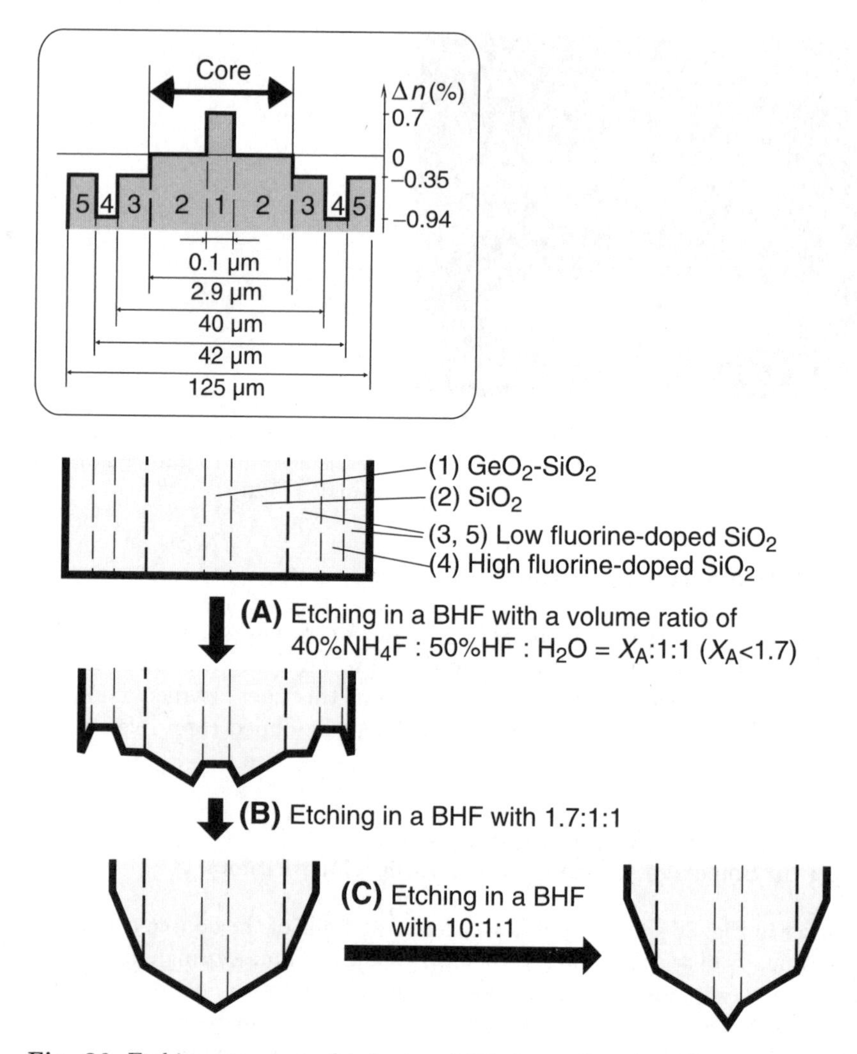

Fig. 36. Etching process to fabricate a triple-tapered probe. The inset shows the cross-sectional profile of the relative refractive index difference Δn of a multistep index fiber produced to fabricate the probe

tapering of sections 2 and 3, and (C) sharpening of section 1. Here, a BHF with a volume ratio of NH_4F solution (40%): HF acid (50%): H_2O is denoted as X:1:1. The values of X in steps A–C are $X_A = 0.6$, $X_B = 1.7$, and $X_C = 10$, respectively.

Figures 37a and b show scanning electron micrographs of the fabricated triple-tapered probe and its top region, respectively. The probe has an apex diameter of less than 10 nm. The values of the base diameters of d_{B1} and d_{B2} are 100 nm and 2 μm, respectively. The first cone angle θ_1, the second

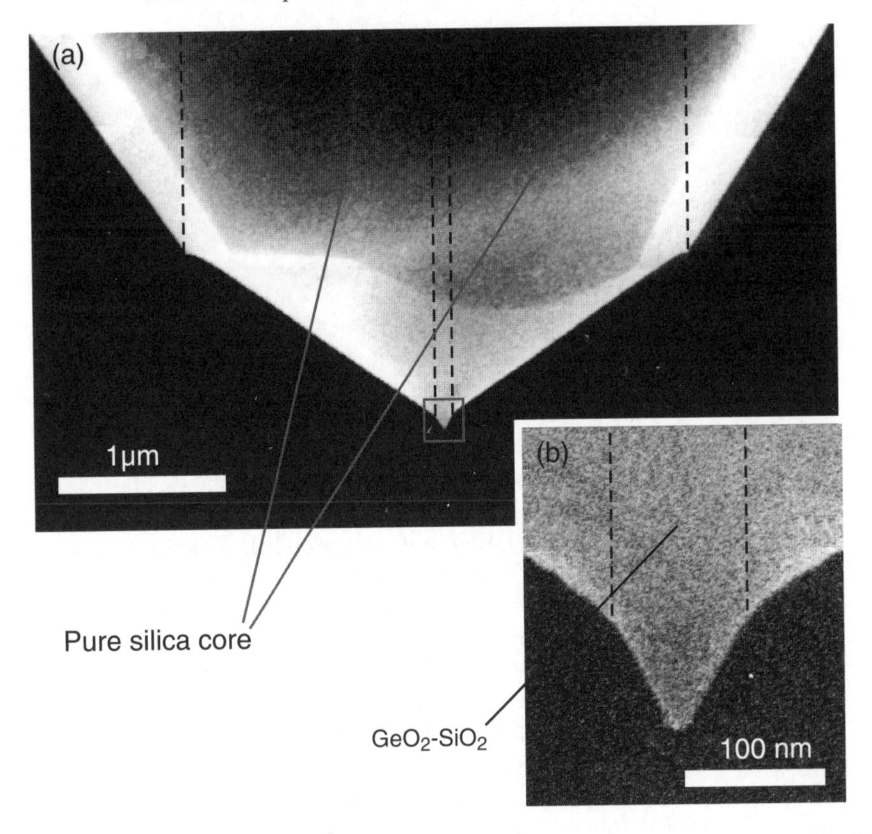

Fig. 37. Cross-sectional profile of the index difference Δn of a multistep index fiber produced to fabricate the probe; SEM micrographs of (**a**) a triple-tapered probe and (**b**) its magnified apex region. $\theta_1 = 60°$; $\theta_2 = 120°$; $\theta_3 = 60°$; $d < 10$ nm

Table 5. Parameters for drawing process to produce the double-core fiber

Drawing speed	150 m/s
Temperature	2106°C
Drawing tension	75 g
Clad diameter of the preform glass rod	25.5 mm
Clad diameter of the drawn fiber	125 µm

cone angle θ_2, and the third cone angle θ_3 are 60° (±1°), 120° (±1°), and 60° (±1°), respectively. Table 6 summarizes the parameters of the etching process for fabricating the probe of Fig. 37a. After the etching, the triple-tapered probe was coated with a 200-nm thick aluminum film by a vacuum evaporation method.

Table 6. Parameters of etching process for fabricating the triple-tapered probe

Step	Volume ratio of BHF	Etching time	Temperature
A	0.6:1:1	40 min	25°C
B	1.7:1:1	20 min	25°C
C	10:1:1	10 min	25°C

7.2 UV Near-Field Photoluminescence Images of Polysilane

To demonstrate the UV triple-tapered probe, we performed UV near-field photoluminescence imaging of polysilanes, which can be regarded as quantum wires based on silicon, and are expected to be applied to UV-light emitting devices. By changing the conformation of the silicon backbones, the optical properties can be controlled. Figures 38a and b shows absorption (dotted curves) and photoluminescence (solid curves) spectra of two polysilanes named polydihexylsilane (PDHS) and n-decyl-(s)-2-methylbutyl silane (Chiral-PS), respectively. They have different backbone structures called transplanar and 7/3 helical, respectively. For the first UV photoluminescence imaging, PDHS was chosen, which has absorption and photoluminescence peaks at 370 nm and 380 nm, respectively. The quantum efficiency is evaluated to be 5%. PDHS was fixed on a glass substrate. Figure 39 shows a schematic diagram of the i-mode UV-SNOM employing the triple-tapered probe. Here, an argon ion laser with a 351 nm wavelength is coupled into the fiber probe, and the UV optical near field generated around the probe excites the sample. Photoluminescence of PDHS is collected by an objective lens or silica ball lens. The triple-tapered probe has three cone angles of 60°, 120°, and 60° as shown in Fig. 40a. Figure 40b shows a UV near-field photo-

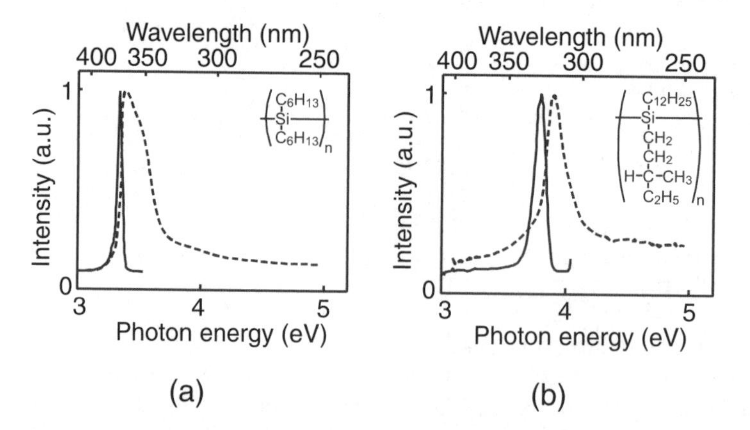

Fig. 38. Absorption (*dotted curves*) and photoluminescence (*solid curves*) spectra of polydihexylsilane (PDHS) and n-decyl-(s)-2-methylbutyl silane (Chiral-PS), respectively. These polysilanes have different backbone structures called transplanar and 7/3 helical, respectively

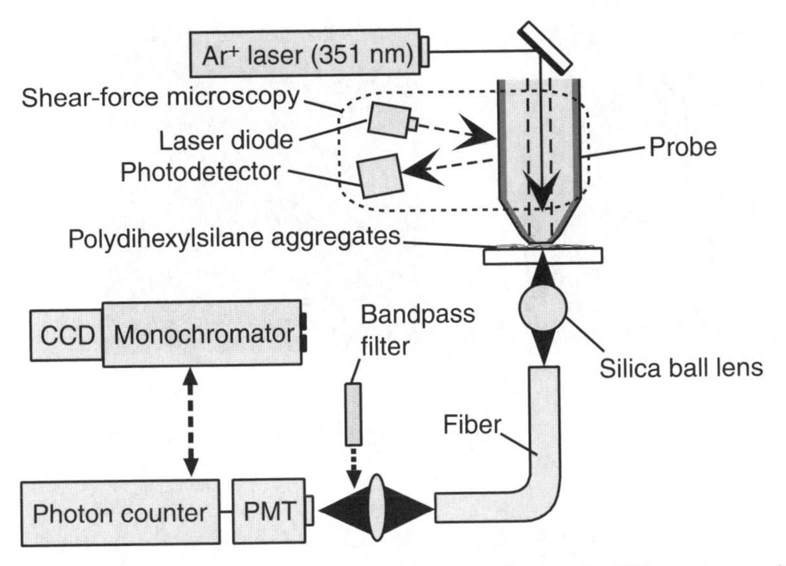

Fig. 39. Schematic diagram of the experimental setup for UV spectroscopic applications. Here, the photon-counting detection is performed at i-mode SNOM imaging. The spectrum of the sample is obtained by using the CCD camera and monochromator

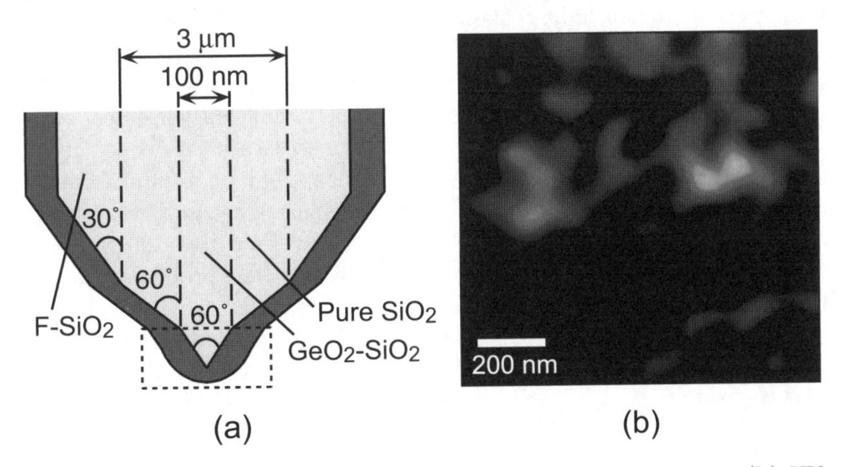

Fig. 40. (**a**) Triple-tapered fiber probe with a pure silica core and (**b**) UV near-fild photoluminescnece image of transplanar-type polydihexylsilane obtained with i-mode SNOM and probe as shown in (**a**) and (**b**), respectively. The top region of the probe, surrounded by the dotted rectangular in (**b**), was removed before the imaging

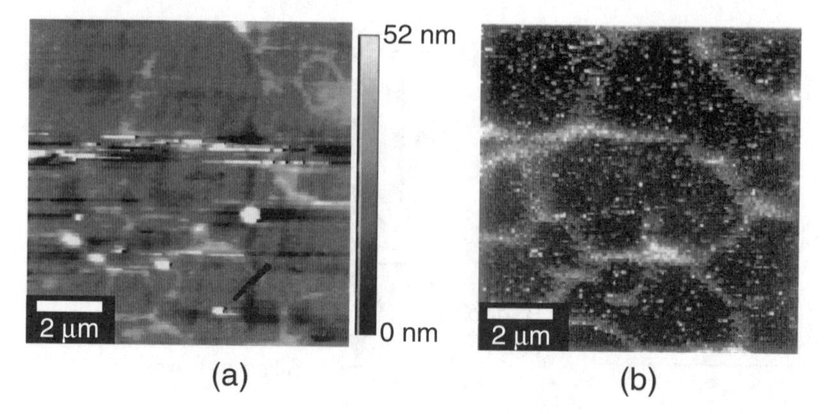

Fig. 41. (**a**) Topographic image and (**b**) i-mode near-field photoluminescence image of chiral-PS molecules at 325 nm

luminescence image of PDHS at 380 nm. The FWHM of the bright portion is estimated to be around 100 nm. Such a near-field signal could not be detected through double-tapered probes with the GeO_2-doped core due to background originated from the luminescence of the core.

Next, by applying the above-described UV-SNOM technique to chiral-PS, we succeeded in obtaining i-mode near-field photoluminescence images of chiral-PS molecules dispersed on ultraflat sapphire substrate. Figure 41a shows a topographic image of chiral-PS. The height of the observed fibril-like structure is 7–50 nm. Since the diameter of Chiral-PS molecules has been estimated to be around 1 nm, we guess the fibril-like structures are constructed of several molecules. Figure 41b shows a UV near-field photoluminescence image of chiral-PS molecules at 325 nm in the same area as Fig. 41a. For the excitation of the sample, UV light with a wavelength of 300 nm was used. The achieved spatial resolution is around 60 nm. The probe used was a UV triple-tapered probe with a small core diameter of around 2 μm. It was fabricated based on selective etching of a new double-core fiber, which was produced by only vapor-phase axial deposition. Details of this fiber will be discussed elsewhere.

Near-Field Raman Spectroscopy of Silicon Crystal

Near-field Raman spectroscopy of silicon is considered a suitable method for analyzing silicon-based electronic devices. The Raman signal of silicon has a rather weaker sattering efficiency and a smaller wave number than that of polydiacetylene, and competes with the background signal of Raman scattering from the GeO_2-doped silica core. To suppress the background, a fiber probe with a pure silica core should be used. The background of the probe is suppressed to less than 10% compared with that of a double-tapered probe.

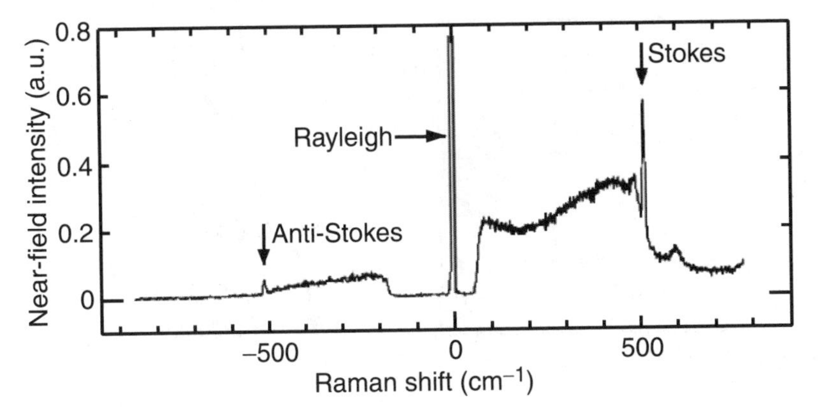

Fig. 42. Near-field Raman spectrum of a silicon crystal obtained by an aliuminized fiber probe with a pure silica core and an aperture diameter of 100 nm

Figure 42 shows a near-field Raman spectrum of a silicon crystal, obtained by i-mode SNOM employing the triple-tapered probe with an aperture diameter of 100 nm. A sharp peak from silicon is clearly seen at 520 cm^{-1}. The broad signal around 400 cm^{-1} corresponds to the Raman scattering from the SiO_2 core. It is found that the fiber probe is very effective in suppressing the background originated from the GeO_2.

7.3 Fabrication of a Pure Silica Fiber Probe by Pulling and Etching

Among commercial single-mode fibers, only 1.3-μm fiber (PSF) with a pure silica core has a high transmittance at near UV. We developed a new method of tapering a 1.3-μm PSF based on a combination of pulling and etching. The method involves two steps: (A) heating-and-pulling the fiber by a micropipette puller and (B) etching the fiber in buffered hydrogen fluoride solution as shown schematically in Fig. 43a. Figure 43b shows the magnified top region of the tapered shape formed by step B. Here, θ and $2r_{1E}$ are the cone angle in the apex region of the fiber and the reduced core diameter at the end of the tapered core, respectively. Figure 43c shows the cross-sectional profile of the tapered probe obtained by increasing the etching time in step B. Details of this profile are discussed later. The fabricated tapered fiber has the tapered shape and the magnified top region as seen in the scanning electron micrographs of Figs. 44a and b, respectively. Figures 44c and d correspond to the magnified view of regions indicated by squares in b and c, respectively. The cone angle and the apex diameter are $\theta = 65°$ and 10–20 nm, respectively. By investigating 20 fiber samples, we obtained 80% reproducibility for a cone angle $\theta = 60°$ ($\pm 5°$) and an apex diameter of 10–20 nm. We now describe the method for producing this pure silica core fiber.

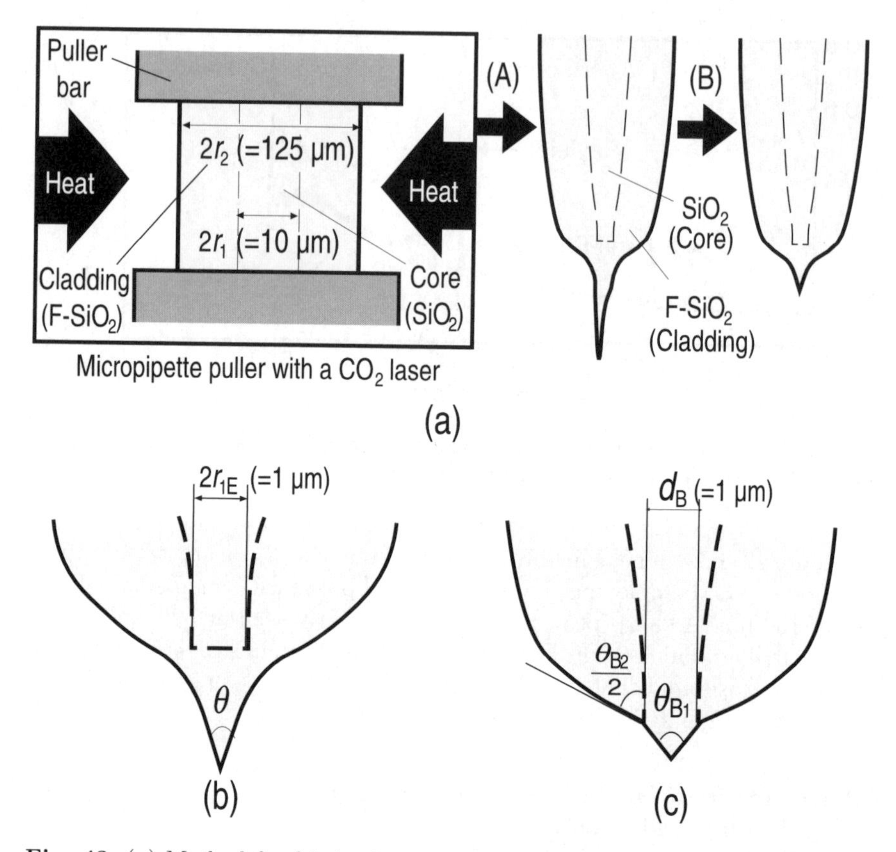

Fig. 43. (**a**) Method for fabricating a tapered fiber with the pure silica core. (**b**) Cross-sectional profile of the top of the fiber formed by step A in (**a**). r_{1E}, reduced core diameter at the end of the tapered core; θ_{B1}, θ_{B2}, cone angles of the core and clad, respectively. (**c**) Cross-sectional profile obtained by increasing the etching time in step B

Table 7. Parameters of the puller for a pure silica core fiber

Heat $\mathcal{H}$, CO_2 laser power	350
Filament $\mathcal{F}$, length of the fiber scanned with the laser beam	0
Velocity $\mathcal{V}$, velocity of the puller bar at the end of the heating time	1
Delay $\mathcal{D}$, delay time between the heating and pulling	130
Pull $\mathcal{P}$, strength of pull	150

In step A, a PSF with core diameter of 10 µm, a clad diameter of 125 µm, and an index difference of −0.3% was heated and pulled by a micropipette puller (Sutter Instrument, P-2000) combined with a CO_2 laser. This puller has been mechanically adjusted to fabricate a micropipette with a diameter of 1 mm. To pull the fiber, the two puller bars were exchanged to optional ones on which a 125-μm bare fiber can be attached. In our case, to produce a probe with a bare-portion length as small as 1 cm, a plastic-coated portion of the fiber sample was carefully attached on one bar, and a bare portion was fixed to another bar. The puller is adjusted with its parameters shown in Table 7. Here, the heat parameter $\mathcal{H}$ ($0 \leq \mathcal{H} \leq 999$) fixes the CO_2 laser power. The filament parameter $\mathcal{F}$ is the length of the fiber that is scanned with the CO_2 laser beam. The velocity parameter $\mathcal{V}$ ($1 \leq \mathcal{V} \leq 255$) shows the velocity of the puller bar at the end of the heating time. The delay parameter $\mathcal{D}$ ($0 \leq \mathcal{D} \leq 255$) represents the delay time between the end of the heating and the beginning of the pulling (in milliseconds). The puller is mechanically adjusted to make the delay zero at $\mathcal{D} = 125$. The pull parameter $\mathcal{P}$ decides the pull strength and is controlled in a region of $0 \leq \mathcal{P} \leq 255$. If step A is repeated using the same puller and the same fiber, some of the parameters, such as $\mathcal{H}$ and $\mathcal{D}$, may be changed to fabricate a pulled fiber as seen in Fig. 44a.

In step B, the fiber was etched by immersing for 10 min in a buffered hydrogen fluoride solution (BHF) with a volume ratio of 40%-NH_4F aqueous solution : 50%-HF acid : deionized water = 1.7:1:1. The temperature of the BHF was 25°C (±0.1°C). The fluorine-doped clad and the pure silica core have dissolution rates of $R_1 = 6.6$ μm h^{-1} and $R_2 = 7.6$ μm h^{-1}, respectively. The pulled and etched probe with a few tens of nanometers was obtained by stopping the etching before the core protruded from the fluoride-doped clad. The cone angle θ is increased by increasing the etching time T. In our experiments, the cone angles were 35° and 65° for etching times of 5 and 10 min, respectively. However, we could not realize the pulled and etched probe with a reproducibility of more than 80% due to the mechanical misalignment of the puller.

On the other hand, once the core is exposed from the fluorine-doped silica, the core is selectively etched due to the difference of the dissolution rate R_1 of the core and the dissolution rate R_2 of the clad ($> R_1$). Then, the fiber has two cone angles of θ_{B1} and θ_{B2}, as shown in Fig. 43c, where d_B is the base diameter of the conical core, with the angle θ_{B1} given by

$$\sin\frac{\theta_{B1}}{2} = \frac{R_1}{R_2}\sin\frac{\theta_{B2}}{2} . \tag{11}$$

The pulled and etched probe as shown in Fig. 21a was fabricated by selective etching and has a GeO_2 tip protruding from the silica clad. In the case of using a PSF, the cross section of the apex region was elliptical, with a longer principal diameter of around 200 nm. Therefore, we could not employ the selectively etched PSF as a probe. The value of d_B is equal to the reduced core diameter $2r_{1E}$ of the fiber as shown in Fig. 43a. The reduced core diameter is estimated to be 2 μm from a SEM image of selectively etched fibers that were immersed in BHF with 1.7:1:1 for times longer than 10 min. It is straightforward to obtain a minimum ratio of $r_{1E}/r_1 = 1/5$.

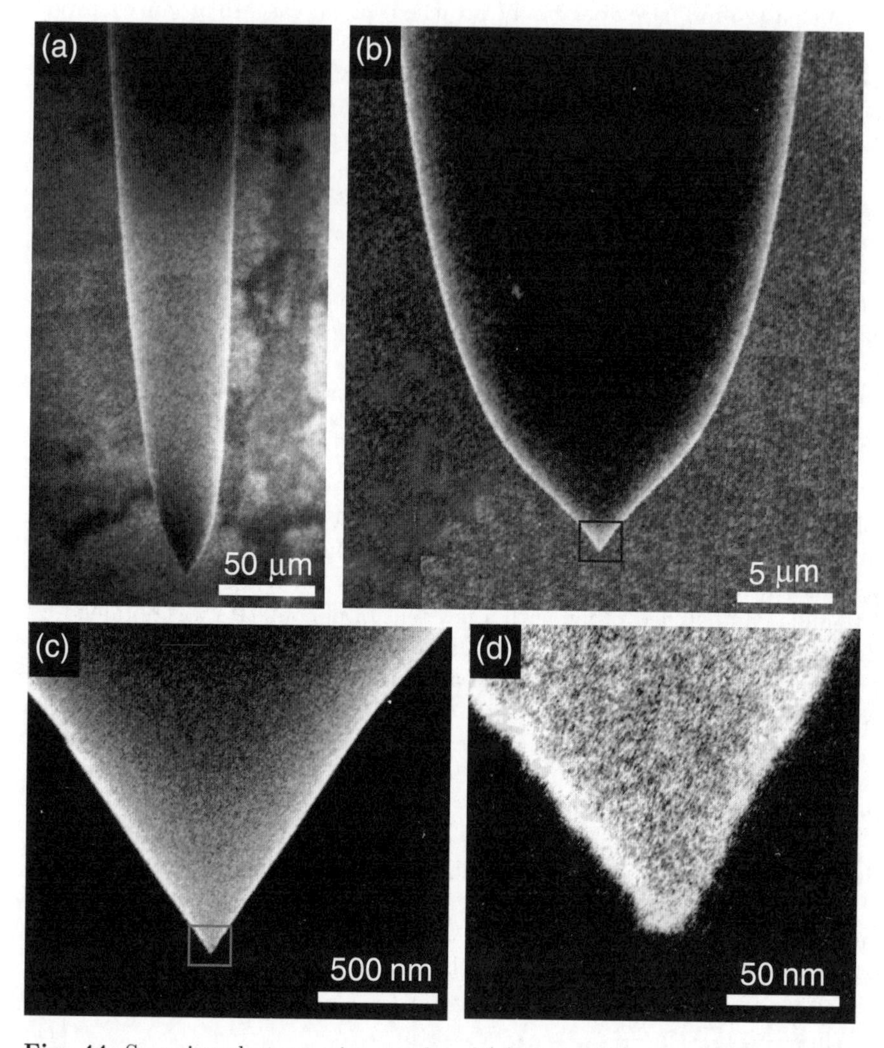

Fig. 44. Scanning electron micrographs of (**a**) the pulled and etched probe with the pure silica core and (**b**) its magnified top region; (**c, d**) The magnified apex regions as indicated by the squares in (**b**) and (**c**), respectively. The cone angle and apex diameter are 65° and 20 nm, respectively

UV Application Performed with i-mode SNOM Employing a Pulled- and Etched-Probe

The pure silica tapered probes were aluminized by a vacuum evaporation method, and applied to a UV near-field photoluminescence image of polydihexylsilane. Figures 45a and b show a shear-force topographic image and near-field photoluminescence image at 380 nm. In Fig. 45b, the value of

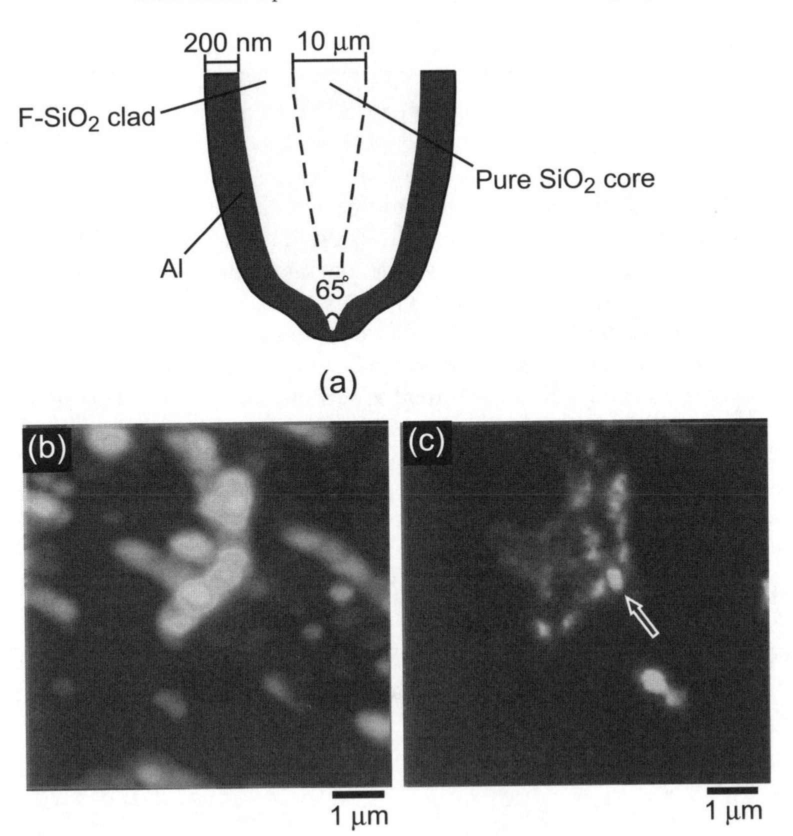

Fig. 45. (**a**) Schematic illustration of the probe fabricated by pulling and etching a commercial fiber with the pure silica core; (**b**) Shear-force topographic image and (**c**) i-mode near-field photoluminescence images of polydihexylsilane aggregates at 380 nm

FWHM of the bright portion as indicated by the arrow is estimated to be around 100 nm. The background signal in the dark region is almost zero, in contrast to the maximum counting rate of the photoluminescence signal. Since the pulled and ethced fiber can be fabricated by pulling and etching commercial fiber, the probe has been widely used as a low-noise probe for near-field fluorescent imaging in the visible region [49] and UV nanofabrication [50] based on photoenhanced chemical vapor deposition.

8 Outlook

The resolution capability and throughput of SNOM have been greatly improved by employing application-oriented probes such as a double-tapered probe with an aperture diameter of a few tens of nanometers and a pure silica fiber probe, and various imaging and spectroscopic applications have been successfully carried out. However, the techniques for mass producing these probes is not yet established due to the low reproducibility of the metallizing process for forming a nanometric aperture. Furthermore, single-mode fiber probes should be newly developed for some imaging applications such as polarization measurements in the ultraviolet and visible regions. If these techniques are established, SNOM and related techniques will be further used widely by many researchers and engineers. To realize the mass production of near-field optical probes, the author has concentrated on the development of a novel metallization technique based on electroless plating [51] and on the fabrication of new single-mode fibers.

Acknowledgments

The author wishes to thank M. Ohtsu, H. Ito, R. Uma Maheswari, T. Saiki, M. Naya, Y. Toda, and N. Hosaka for their assistance and fruitful discussions.

Appendix

Fabrication of Tapered Fibers Based on Hybrid Selective Etching of a Double-Clad Fiber

The double-clad fiber [22] was developed for fabricating pencil-shaped probes and triple-tapered probes. These probes can be fabricated with almost 100% reproducibility based on selective etching. In this section, we describe the structure of a triple-tapered probe and design/fabrication of the fiber.

Triple-Tapered Probe

Figures 46a and b show cross-sectional profiles of the single-tapered probe and triple-tapered probe. In Fig. 46a, θ is the cone angle and λ is the optical wavelength in the fiber. In Fig. 46b, L is defined as the length of the portion with a cross-sectional diameter of λ to the apex. L_1 is the length of the first taper, θ_1, θ_2, and θ_3 are the cone angles of the first, second, and third tapers, respectively, and d_{B1} and d_{B2} are the base diameters of the first and second tapers, respectively. As described in Sect. 3.6, the light entering the single-tapered probe is strongly attenuated by metal from the portion with diameter λ to the apex. To reduce the attenuation, the length L must be decreased by

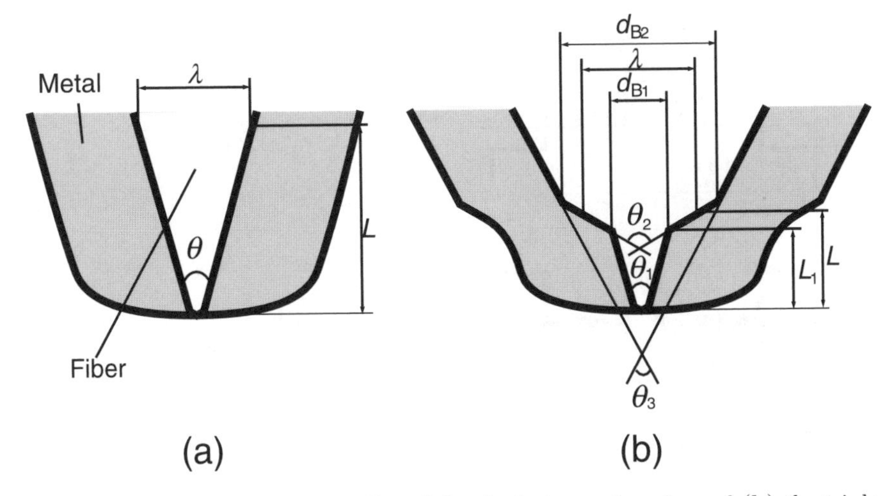

Fig. 46. (**a**) Cross-sectional profiles of the single-tapered probe and (**b**) the triple-tapered probe; θ, cone angle of the tapered core; λ, optical wavelength; L, length of the tapered core with the foot diameter equal to λ; t_S, skin depth of the metal; θ_1, θ_2, θ_3, cone angles of the first, second, and third tapers, respectively; d_{B1}, d_{B2}, base diameters of the first and second tapers, respectively; L_1, length of the first taper

increasing the cone angle. In the case of fabricating the tapered probe with a large cone angle, i.e., 120°, one must accept the limited resolution affected by optical leaking from the metal around the apex region of the probe. In the triple-tapered probe, we can decrease the length L by increasing θ_2 in order to enhance the throughput. Thus, the resolution capability is increased by simultaneously decreasing the cone angle and decreasing the first taper length L_1 to a few hundreds of nanometers, which corresponds to several times the skin depth of metal.

Selective Etching of a Double-Clad Fiber

Figure 47a shows the cross-sectional profile of the index difference of the double-clad fiber. Here, the fiber involves three sections: (1) a GeO_2-doped silica core, (2) a pure silica clad, and (3) a fluorine-doped silica support. The index differences of sections 1 and 3 with respect to section 2 are 1.2% and −0.7%, respectively. The radii of sections 1–3 are $r_1 = 0.65$ µm, $r_2 = 13.5$ µm, and $r_3 = 62.5$ µm, respectively. Figure 47b shows a schematic diagram of the etching process consisting of three steps A–C. By defining the dissolution rate of section i in step j as R_{ij} (where i = 1, 2, 3; j=A, B, C), the relative dissolution rates R_{ij}/R_{3j} is represented by Fig. 47c. As described in Sect. 2.3, the relative dissolution rate of GeO_2-doped glass to pure silica glass depends on the concentration of buffered HF mixed with a volume ratio of 40%NH_4F sol. : 50%HF acid: H_2O = X:1:Y. We now use 1.7:1:1, 10:1:1, and 1.7:1:5, in

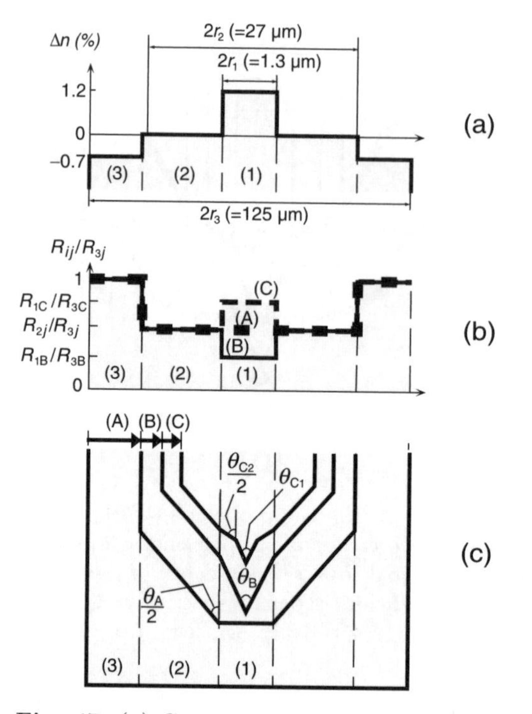

Fig. 47. (**a**) Cross-sectional profile of the index difference Δn. r_1, radius of the GeO_2-doped silica core; r_2, radius of the pure silica clad; r_3, radius of a fluorine-doped silica support; (**b**) Etching method for fabricating application-oriented probes; (**c**) Cross-sectional profile of the relative dissolution rate

which the dissolution rates of sections 1 and 2 are experimentally found to satisfy the relation

$$\frac{R_{1B}}{R_{2B}} = 0.29 < \frac{R_{1A}}{R_{2A}} = 1.0 < \frac{R_{1C}}{R_{2C}} = 1.48\,. \tag{12}$$

On the other hand, the relative dissolution rate of the fluorine-doped section to the pure silica section is approximately fixed for the concentration of BHF. So,

$$R_{2j}/R_{3j} = 0.51 \qquad (j = \mathrm{A},\ \mathrm{B},\ \mathrm{C})\,. \tag{13}$$

We now discuss the etching process using 1.7:1:1, 10:1:1, and 1.7:1:5 in steps A–C, respectively.

In step A, the fiber is tapered to an angle of θ_A represented by

$$\sin(\theta_A/2) = R_{2j}/R_{3j} \qquad (\text{where} \quad j = \mathrm{A},\ \mathrm{B},\ \mathrm{C})\,. \tag{14}$$

If the fiber diameter is equal to $[2r_2]$ after step A, the etching time T_A is represented by

$$T_{\mathrm{A}} = (r_3 - r_2)/R_{3\mathrm{A}} \,. \tag{15}$$

The tapered fiber will have an apex diameter smaller than $2r_1$ if $T_{\mathrm{A}} \geq \tau_{\mathrm{A}}$, where τ_{A} is the time required to make an apex diameter of $2r_1$ and is

$$\tau_{\mathrm{A}} = \frac{r_2 - r_1}{R_{3\mathrm{A}}} \sqrt{\frac{R_{2\mathrm{A}} + R_{3\mathrm{A}}}{R_{3\mathrm{A}} - R_{2\mathrm{A}}}} \,. \tag{16}$$

Thus, it is straightforward to find that the radius r_2 of section 2 must be smaller than the critical radius $r_{2\mathrm{P}}$ expressed as

$$r_{2\mathrm{P}} = \frac{r_3 + \xi \cdot r_1}{1 + \xi}, \qquad \text{where} \quad \xi = \sqrt{\frac{R_{2\mathrm{A}}/R_{3\mathrm{A}} + 1}{1 - R_{2\mathrm{A}}/R_{3\mathrm{A}}}} \,. \tag{17}$$

In step B, section 1 is sharpened with a different angle θ_{B}, given by

$$\sin(\theta_{\mathrm{B}}/2) = R_{1\mathrm{B}}/R_{3\mathrm{B}} \,. \tag{18}$$

We obtain pencil-shaped probes with zero apex diameter and cone angles θ_{A} and θ_{B} when the etching time T_{B} is larger than τ_{B}, expressed as

$$\tau_{\mathrm{B}} = \frac{r_1}{R_{1\mathrm{B}}} \sqrt{\frac{R_{1\mathrm{B}} + R_{3\mathrm{B}}}{R_{3\mathrm{B}} - R_{1\mathrm{B}}}} \,. \tag{19}$$

Furthermore, to obtain a triple-tapered probe, we perform step C, when the largest cone angle θ_{C2} is given by

$$\sin\frac{\theta_{\mathrm{C2}}}{2} = \frac{R_{1\mathrm{C}}}{R_{3\mathrm{C}}} \qquad (\text{where} \quad R_{1\mathrm{C}} > R_{2\mathrm{C}}) \,. \tag{20}$$

The cone angle θ_{C1} is increased from θ_{B} by increasing the etching time T_{C}, and is equal to θ_{C2} at $T_{\mathrm{C}} > \tau_{\mathrm{C}}$, where τ_{C} is given by

$$\tau_{\mathrm{C}} = \frac{r_1}{R_{1\mathrm{C}}} \sqrt{\frac{R_{1\mathrm{C}} + R_{3\mathrm{C}}}{R_{3\mathrm{C}} - R_{1\mathrm{C}}}} \,. \tag{21}$$

Therefore, we can obtain a triple-tapered probe and a pencil-shaped probe with a cone angle $\theta_{\mathrm{B}} = \theta_{\mathrm{C2}}$ at $0 < T_{\mathrm{C}} < \tau_{\mathrm{C}}$ and $T_{\mathrm{C}} > \tau_{\mathrm{C}}$, respectively.

To obtain a cone angle θ_{A} as small as $62°$, the index difference of section 3 is estimated from (15) and (19) to be as high as 0.7%. To obtain a cutoff wavelength of around 400 nm, section 1 is tailored with an index difference of 1.2% and a core radius of $r_1 = 0.65$ µm. Then we obtain an estimated value of $\theta_{\mathrm{A}} = 17°$ from (13) and (19). Furthermore, when the clad radius r_3 is a standard value of 62.5 µm, we obtain a critical radius of $r_{2\mathrm{P}} = 23$ µm from (14) and (18). We make the radius r_2 have a value of 13.5 µm, which is smaller than the critical radius.

To realize the fiber as shown in Fig. 47a, we fabricated a preform glass rod by vapor-phase axial deposition and drew the fiber. To suppress the

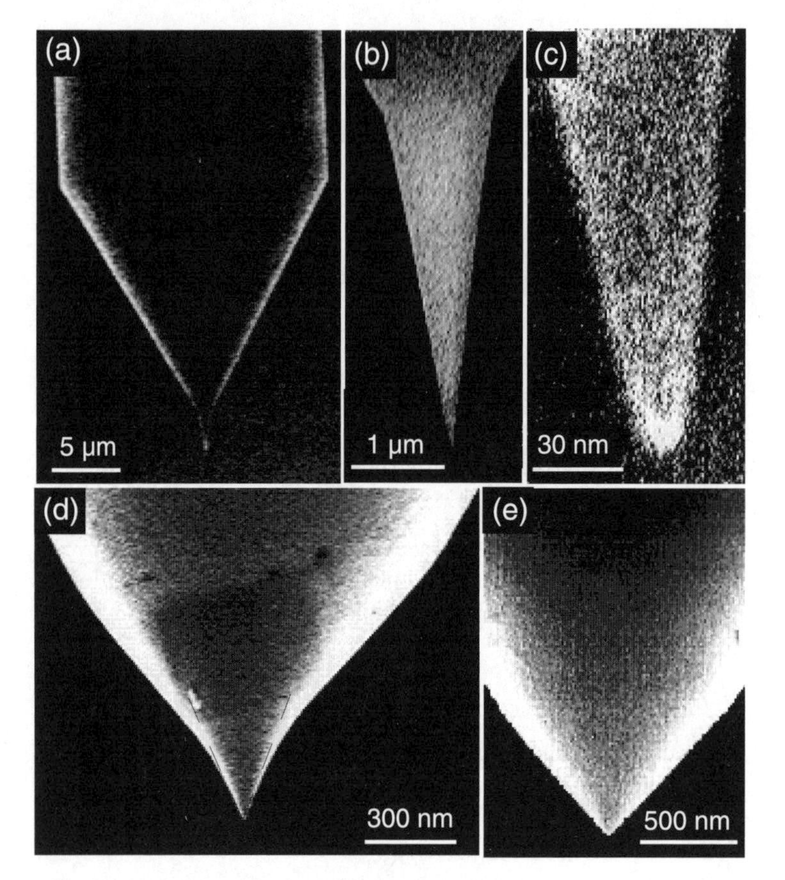

Fig. 48. (**a–c**) SEM micrographs of a pencil-shaped probe, its magnified tapered core, and its magnified apex region. $\theta_\mathrm{A} = 62°$; $\theta_\mathrm{B} = 17°$; $d < 10$ nm; (**d**) SEM micrograph of a triple-tapered probe. $\theta_\mathrm{A} = 62°$; $\theta_\mathrm{C1} = 50°$; $\theta_\mathrm{C2} = 85°$. (**e**) SEM micrograph of a pencil-shaped probe. $\theta_\mathrm{A} = 62°$; $\theta_\mathrm{B} = 85°$

diffusion of GeO_2 and fluorine, the drawing tension should be as high as possible. However, in the case of drawing the fiber with a high tension of 60 g, we could not reproducibly cleave the fiber to obtain a flat facet with a commercial fiber cleaver. We consider that the low reproducibility can be attributed to the remaining stress between sections 2 and 3. To suppress this remained stress, we kept a low tension of less than 28 g during the drawing.

To demonstrate the tailoring capability of the different types of probes, we performed the etching process using the fabricated fiber. We prepared 30 fiber samples with flat ends. The fibers were etched consecutively for $T_\mathrm{A} = 40$ min in a 1.7:1:1 solution and for $T_\mathrm{B} = 20$ min in a 10:1:1 solution. We obtained a pencil-shaped probe with a small cone angle for high resolution. Figures 48a–c show SEM micrographs of the probe, the magnified tapered core, and the

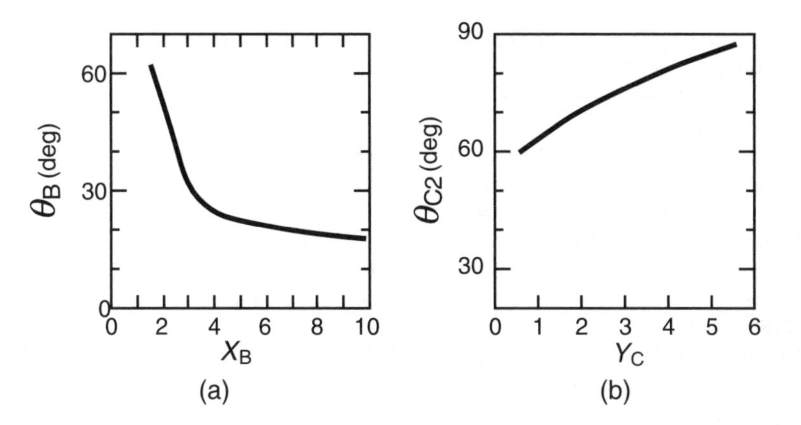

Fig. 49. (**a**) Dependencies of the taper angle θ_B on the volume ratio X_B of NH_4F aqueous solution in the etching solution; (**b**) Dependency of the taper angle θ_{C2} on the volume ratio of H_2O in the etching solution

magnified apex region, respectively. Here, the apex diameter is less than 10 nm. The cone angles are $\theta_A = 62°$ and $\theta_B = 17°$, which agree with the estimated values from (15) and (19).

Furthermore, by etching the pencil-shaped probe in a 1.7:1:5 solution for $T_C = 2$ min, we obtained a triple-tapered probe with a high-resolution capability and high throughput. Figure 48d shows a SEM micrograph of the triple-tapered probe. The probe has three cone angles of $\theta_{C1} = 50°$, $\theta_{C2} = 85°$, and $\theta_A = 62°$ and an apex diameter of less than 10 nm. The two base diameters of tapers with $\theta_{C1} = 50°$ and $\theta_{C2} = 85°$ are $d_{B1} = 250$ nm and $d_{B2} = 1.28$ μm, respectively. At $T_C = 2.75$ min, we obtained another type of pencil-shaped probe with a large angle of 85° near the apex region as shown in Fig. 48e. Figures 49a and b show the dependencies of θ_B and θ_{C2}, respectively, on the volume ratios of BHF. Here, 1.7:1:1, X_B:1:1, and 1.7:1:Y_C are used in steps A–C. The cone angles θ_B and θ_{C2} are controlled by varying X_B and Y_C, respectively.

References

1. E. Betzig, P.L. Finn, J. S. Weiner: Appl. Phys. Lett. **60**, 2484 (1992)
2. R. Toledo-Crow, P.C. Yang, Y. Chen, Vaez-Iravani: Appl. Phys. Lett. **60**, 2957 (1992)
3. D.W. Pohl, W. Denk, M. Lanz: Appl. Phys. Lett. **44**, 651 (1984)
4. E. Betzig, M. Isaacson, A. Lewis: Appl. Phys. Lett. **51**, 2088 (1987)
5. D.R. Turner: US Patent, 4,469,554 (1983)
6. D. Courjon, K. Sarayeddine, M. Spajer: Opt. Commun. **71** 23 (1989)
7. R.C. Reddick, R.J. Warmack, T L. Ferrel: Phys. Rev. B **39** 767 (1989)
8. K.M. Takahashi: J. Colloid Interface Sci. **134**, 181 (1990)

9. T. Hartmann, R. Gatz, W. Wiegräbe, A. Kramer, A. Hillebrand, K. Lieberman, W. Baumeister, R. Guckenberger: 'A Scanning Near-Field Optical Microscope (SNOM) for Biological Applications', In: *Near Field Optics*, **242**, NATO ASI series E, (Kluwer, Dordrecht 1993) 35
10. P. Hoffmann, B. Dutoit, R.-P. Salathé: Ultramicroscopy **61**, 165 (1995)
11. S. Mononobe, R. Uma Maheswari, M. Ohtsu: Opt. Express **1**, 229 (1997), http://epubs.osa.org/opticsexpress/
12. H. Muramatsu, N. Chiba, N. Yamamoto, K. Honma, T. Ataka, M. Shigeno, H. Monobe, M. Fujihira: Ultramicroscopy **71**, 73 (1998)
13. P. Lambelet, A. Sayah, M. Pfeffer, C. Philipona, F. Marquis-Weible: Appl. Opt. **37**, 7289 (1998)
14. E. Betzig: 'Principles and Application of Near-Field Scanning Optical Microscopy (NSOM)', In: *Near Field Optics*, **242**, NATO ASI series E (Kluwer, Dordrecht 1993) 7
15. E. Betzig, J.K. Trautman: Science **257**, 189 (1992)
16. B.I. Yakobson, A. LaRosa, H.D. Hallen, M.A. Paesler: Ultramicroscopy **61**, 179 (1995)
17. G.A. Valaskovic, M. Holton, G.H. Morrison: Appl. Opt. **34**, 1215 (1995)
18. M. Garcia-Parajo, T. Tate, Y. Chen: Ultramicroscopy **61**, 155 (1995)
19. S. Jiang, H. Ohsawa, K. Yamada, T. Pangaribuan, M. Ohtsu, K. Imai, A. Ikai: Jpn. J. Appl. Phys. **31**, 2282 (1992)
20. R. Uma Maheswari, S. Mononobe, M. Ohtsu: J. Lightwave Technol. **13**, 2308 (1995)
21. P. Tomanek: 'Fiber Tips for Reflection Scanning Near-Field Optical Microscopy', In: *Near Field Optics*, **242**, NATO ASI series E, (Kluwer, Dordrecht, 1993) 295
22. S. Mononobe, M. Ohtsu: IEEE Photonics Technol. Lett. **10**, 99 (1998)
23. S. Mononobe, T. Saiki, T. Suzuki, S. Koshihara, M. Ohtsu: Opt. Commun. **146**, 45 (1998)
24. M. Spajor, A. Jalocha: 'The Reflection Near Field Optical Microscope: An Alternative to STOM', In: *Near Field Optics*, **242**, NATO ASI series E, (Kluwer, Dordrecht 1993) 87
25. S. Mononobe: 'Probe fabrication', In: *Near-Field Nano/Atom Optics and Technology* (Springer, Berlin 1998) 31
26. S. Mononobe, M. Ohtsu: J. Lightwave Technol. **14**, 2231 (1996); Erratum J. Lightwave Technol. **15**, 162 (1997)
27. S. Mononobe, M. Ohtsu: J. Lightwave Technol. **15**, 1051 (1997)
28. E. Betzig, J.K. Trautman, T.D. Harris, J.S. Weiner, Kostelak: Science **251**, 1468 (1991)
29. M. Naya, S. Mononobe, R. Uma Maheswari, T. Saiki, M. Ohtsu: Opt. Commun. **124**, 9 (1996)
30. M. Naya, R. Micheletto, S. Mononobe, R. Uma Maheswari, M. Ohtsu: Appl. Opt. **36**, 1681 (1997)
31. A.V. Zvyagin, J.D. White, M. Ohtsu: Opt. Lett. **22**, 955 (1997)
32. Y. Toda, M. Kourogi, M. Ohtsu, Y. Nagamune, Y. Arakawa: Appl. Phys. Lett. **69**, 827 (1996)
33. R. Uma Maheswari, S. Mononobe, K. Yoshida, M. Yoshimoto, M. Ohtsu: Jpn. J. Appl. Phys. **38**, 6713 (1999)
34. R. Uma Maheswari, H. Tatsumi, Y. Katayama, M. Ohtsu: Opt. Commun. **120**, 325 (1995)

35. N. Hosaka, T. Saiki: J. Microscopy **202**, Part 2, 362 (2001).
36. T. Saiki, S. Mononobe, M. Ohtsu, N. Saito, J. Kusano: Appl. Phys. Lett. **67**, 2191 (1995)
37. Y. Narita, T. Tadokoro, T. Ikeda, T. Saiki, S. Mononobe, M. Ohtsu: Appl. Spectros. **52**, 1141 (1998)
38. M. Arai, S. Koshihara, M. Ueda, M. Yoshimoto, T. Saiki, S. Mononobe, M. Ohtsu, T. Miyazawa, M. Kira: J. Lumines. **87–89**, 951 (2000)
39. T. Saiki: 'Diagnosing Semiconductor Nano-Materials and Devices', In: *Near-Field Nano/Atom Optics and Technology* (Springer, Berlin 1998) 153
40. T. Izawa, S. Sudo: *Optical Fibers: Materials and Fabrication* (KTK Scientific, Tokyo, 1987)
41. S. Mononobe, M. Naya, T. Saiki, M. Ohtsu: Appl. Opt. **36**, 1496 (1997)
42. L. Novotony, D. W. Pohl, B. Hecht: Ultramicroscopy **61**, 1 (1995)
43. K. Yoshida, M. Yoshimoto, K. Sasaki, T. Ohnishi, T. Ushiki, J. Hitomi, S. Yamamoto, M. Sigeno: Biophy. J. **74**, 1654 (1998)
44. T. Saiki, S. Mononobe, M. Ohtsu, N. Saito, J. Kusano: Appl. Phys. Lett. **68**, 2612 (1996)
45. H. Nakamura, T. Sato, H. Kanbe, K. Sawada, T. Saiki: J. Microscopy **202**, 50 (2001)
46. N. Saito, M. Yamaga, F. Sato, I. Fujimoto, M. Inai, T. Yamamoto, T. Watanabe: Inst. Phys. Conf. Ser. **136**, 601 (1993)
47. T. Saiki: 'Nano-Optical Imaging and Spectroscopy of Single Quantum Constituents'. In: *Progress in Nano-Electro-Optics II*, ed. by M. Ohtsu (Springer, Berlin 2003) 111
48. S. Mononobe, T. Saiki, T. Suzuki, S. Koshihara, M. Ohtsu: Opt. Commun. **126**, 45 (1998)
49. H. Aoki, Y. Sakurai, S. Ito, T. Nakagawa: J. Phys. Chem. B **103**, 10553 (1999)
50. V.V. Polonski, Y. Yamamoto, M. Kourogi, H. Fukuda, M. Ohtsu: J. Microscopy **194**, 545 (1999)
51. S. Mononobe, Y. Saito, M. Ohtsu, H. Honma: Jpn. J. Appl. Phys. **43**, (2004), in press

A Novel Method for Forming Uniform Surface-Adsorbed Metal Particles and Development of a Localized Surface-Plasmon Resonance Sensor

H. Takei and M. Himmelhaus

1 A General Method for Preparing Surface-Bound Metal Particles

1.1 Surface-Bound Metal Particles

There is a vast number of studies directed toward surface-adsorbed metal particles, of gold and silver in particular. In contrast to isolated spherical metal particles, analysis of surface-adsorbed particles is made complex because of the interaction between the particle and the solid substrate [1] as well as the interaction among adjacent particles. Moreover, the process of forming metal particles on a substrate quite often leads to deformation of the particle as well as loss of monodispersity in the size [2, 3]. Despite the daunting complexity, surface-bound metal particles continue to receive attention, both experimentally and theoretically, because of a number of fascinating applications. These range from simple applications such as optical filter/absorber and substrates for surface-enhanced spectroscopy to more futuristic photonic applications relying on precise arrangement and interactions among patterned particles [4]. In order to fully develop and exploit these applications, it goes without saying that further progress on both experimental and theoretical fronts is necessary.

In this chapter, we first introduce a novel way of producing surface-adsorbed metal particles; with this method, while uniform samples can be readily formed, a number of important physical parameters can also be controlled precisely, allowing preparation of samples with various optical characteristics. We also mention how these particles can be patterned, which might prove important in regulating interparticle interactions in the future. In the second half of the chapter, we describe a novel optical biosensor that is based on localized surface plasmons associated with the sample so prepared. We will give some measurement examples, followed by general characterization of the sensor system. We then suggest how the unique feature of our sensor can be exploited to make contributions to the field of life sciences, culminating in productive interactions between nano-optics and biotechnology.

1.2 Use of a Monolayer of Monodisperse Dielectric Spheres as a Template

Traditionally, metal particles are formed on a solid substrate by evaporating a minute amount of metal; when the deposited amount is limited, a granular metal film is formed instead of a continuous film [2, 5, 6]. In the case of gold, if a granular film is thermally annealed, gold becomes more or less spherical in shape due to its high surface energy. Because of its simplicity, this system has been widely investigated, but poor control has been a serious impediment. Alternatively, chemically synthesized gold particles can be attached to a substrate [7, 8]; high monodispersity in size can be readily accomplished, but quite often coagulation in the process of surface adsorption can be a serious problem, making it particularly difficult to produce large surface area samples. Either way, it is generally not possible to form well-controlled surface-adsorbed particles without resorting to special exotic techniques such as electron-beam technology [9].

We have devised a novel method for forming uniform surface-adsorbed metal particles at high density. The method originated from our observation that by carefully controlling repulsive forces among suspended polymer nanoparticles and bringing them into contact with a gold substrate [10, 11], it is possible to form a random but uniform monolayer of nanoparticles on the substrate. Such a monolayer can serve as a template for formation of metal particles; upon evaporation, the top half of individual polymer nanoparticles becomes coated with the metal. This is one approach that allows us to form a completely new class of surface-adsorbed metal particles; these particles are cap shaped, and by exploiting the large selection of monodisperse spheres available commercially, it is possible to form samples of different diameters endowed with various optical properties. The deposition thickness also affects the behavior of individual particles. Controlling the average gap distance between adjacent particles alters the collective properties. Here is an opportunity to form novel classes of particles with various optical properties.

So called latex spheres, polymer spheres, have continued to fascinate researchers because of the unparalleled monodispersity. In recent years, it is in the field of photonic bandgap materials that researchers have tried to exploit monodispersity; a two dimensional photonic bandgap material can be obtained by forming a regular layer of spheres through a number of different techniques [4, 12–14]. One can use a technique of raising a substrate out of a solution containing spheres. By careful control of the lifting rate, spheres self-assemble on the substrate, resulting in a regular array. Such an array can be formed also by a flowing sphere suspension against a shallow edge. There are many other means to form a regular array, but reproducible preparation of macroscopic samples with high uniformity remains difficult.

While all these methods lead to many interesting results, one simple method, with potentially many new applications, has not been pursued. This is what we present in this chapter. Our method starts, as with the example presented immediately above, with formation of a monolayer of spheres on a substrate. This monolayer then serves as a template for subsequent evaporation of metals. Evaporated metal adheres only on the side of the sphere facing the evaporation source so that the shape is characterized as a cap [15]. Moreover, the monolayer serves both as an insulator or a spacer. Thus, by using this simple method, we end up with a large number of cap-shaped metal particles that are, electrically speaking, floating over the substrate. By capitalizing on excellent monodispersity, it is easy to envision that a large variety of samples can be readily formed. Interestingly, there are many reports on the use of surface-adsorbed spheres as a temporary mask [16–18], but little attention has been paid to the optical property with the spheres still on the surface so far.

We should refer to some earlier examples of related studies. Fischer and Pohl [19] evaporated gold on top of sparse surface-adsorbed microspheres to form a metal point to study the near-field effect in its vicinity. Haginoya et al. [16] have formed a regular array of microspheres, which served as a mask for subsequent evaporation; here, because the final object of interest was the metal evaporated on the substrate, the spheres were subsequently removed from the substrate. While not dealing with surface-adsorbed particles, polystyrene spheres coated with silver have nonetheless been prepared [20]. We should also mention that use of porous materials as a template for synthesis is becoming popular [21].

1.3 Formation of the Monolayer

The usual method for stabilizing colloidal particles is to chemically tailor their surface in such a way that the repulsive force among the spheres prevents them from sticking together. By gradually decreasing the repulsive force, for example, through an increase in the salt concentration, in the presence of a high-energy surface, spheres begin to adsorb on the surface, forming a monolayer. This is the basis of our methodology [10]. This simple scheme can be expanded along a number of different themes.

We have purchased monodisperse polystyrene spheres, from Polysciences, Warrington, PA, USA (Polystyrene bead Select Certified Size Standard Microsphere) or Dyno Particles, Norway (Dynospheres) and used them as received. For most experiments we used various polystyrene (PS) labware for life-science experiments as substrates. These were purchased from Nunc, Nalgene etc. and used without washing. Prior to forming a monolayer of polystyrene spheres, the PS substrate was coated with 20 nm gold in a vacuum evaporator (Type EBV-6 DA, ULVAC, Kanagawa, Japan). Gold, purity of 99.95%, was purchased from Furuya Metal, Tokyo, Japan and was evaporated from a tungsten boat at a base pressure of 2×10^{-6} Torr at an evaporation rate of 0.1 nm s^{-1}; PS substrates were held at least 30 cm away from the tungsten boat in order to prevent melting of the surface. PS was chosen as the substrate material because it is optically clear and flat PS pieces are readily available. As will be discussed later, microscopic irregularities causing visibly noticeable scattering have to be avoided, but slight warping on a macroscopic scale, as typically happens with molded plastic, could be tolerated quite well. Moreover, it was found that gold evaporated on PS showed much greater adhesion than that on a glass substrate, saving us the extra step of introducing a buffer layer to enhance gold adhesion on; we also noticed that gold evaporated on a PS substrate was significantly free of granules typically formed when gold is evaporated on a glass substrate. PS was obviously not the only suitable polymer; polycarbonate and PMMA apparently serve as the substrate well.

To form a monolayer of PS spheres on a gold-coated PS substrate, partial aggregation of the PS spheres was induced with addition of carbodiimide, 1-ethyl-3-(3-(dimethylamino)propyl) carbodiimide (EDC), Dojindo Laboratories, Kumamoto, Japan, to the PS sphere suspension. While the exact concentration of carbodiimide was a deciding factor on the optical property of the final sample, the concentration is typically in the 10 mM range. As soon as the mixture of the PS spheres and carbodiimide was prepared, the suspension was placed on a gold-coated PS substrate; spheres begin to preferentially adsorb on the substrate, and within a minute, the surface coverage is almost complete. The role played by carbodiimide is fundamentally not chemical in nature, and simple salts such as NaCl and KCl have been found to have a similar effect. It will be explained later, however, that carbodiimide exhibits some characteristics not found with simple salts. To assure reproducibility, the incubation was allowed to proceed for ten minutes, after which nonadsorbed PS spheres were washed off with a copious amount of distilled water, leaving only a monolayer on the substrate. While the sample preparation procedure, summarized in Fig. 1, is rather straightforward, extra care was needed during drying to obtain optically uniform samples. After the sample was dried completely, gold was further evaporated to varying thicknesses.

Sample Characterization

For optical characterization, we used various types of spectrometers including a Hitachi U–3400 (Tokyo, Japan), an Otsuka Electronics Photal IMUC–7000 (Osaka, Japan) and an Ocean Optics, S2000 (Florida, USA). The Hitachi spectrometer was equipped with an integrating sphere. IMUC–7000 and S2000 spectrometers were equipped with CCD linear arrays, allowing instantaneous measurement. The S2000 in particular was configured with a custom-ordered software so that the spectrum could be fitted nonlinearly with a pseudo-Voigt function to give the peak wavelength in real time; the peak wavelength can be displayed as a function of time and stored for later analysis. For electron microscopic characterization, a scanning electron microscope (SEM), Hitachi SEM Model S–800 and a transmission electron microscope (TEM), Hitachi TEH Model HF–2000, were employed.

Figure 2 shows a scanning electron micrograph of a typical sample. Although spheres are randomly adsorbed, the overall uniformity is rather note-

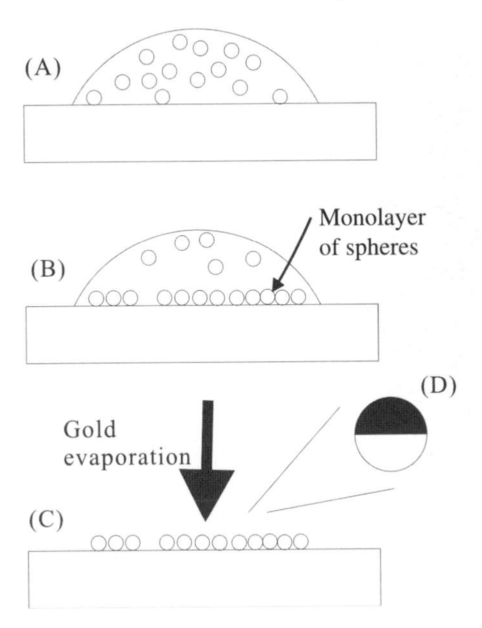

Fig. 1. Schematic diagram showing the procedure for formation of surface-adsorbed, cap-shaped metal particles. (**a**) Suspension of a sphere/carbodiimide mixture placed on a substrate. (**b**) Formation of a monolayer of spheres. (**c**) Evaporation of a metal on top of the monolayer of spheres. (**d**) Cap-shaped metal particle

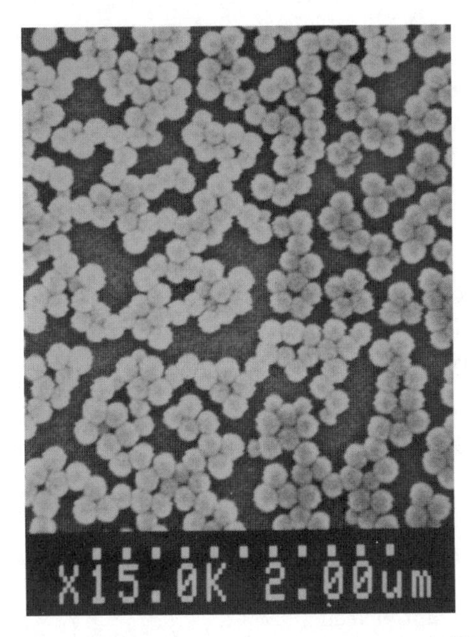

Fig. 2. Scanning electron microscope (SEM) micrograph of a surface-adsorbed sphere monolayer (sphere diameter 110 nm)

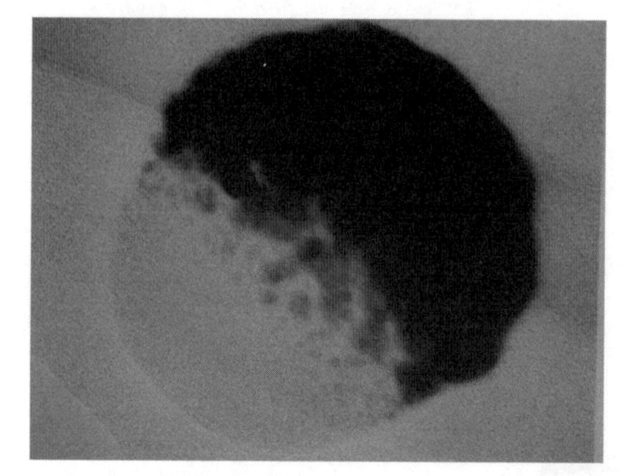

Fig. 3. Transmission electron microscope (TEM) microgram of a dislodged sphere, with twenty nanometers of gold covering one half of a sphere (sphere diameter 110 nm). Surface-adsorbed spheres were dislodged from the substrate by application of ultrasound with a bench-top sonicator. Dislodged spheres were placed on a substrate and observed with TEM, at courtesy of T. Matsumoto

worthy and extends to the macroscopic scale. If a critical carbodiimide concentration is exceeded, however, additional layers begin to form on top of the first layer. As we have found that formation of the second or more layer has a detrimental effect on the absorption spectrum of the final sample [11]; while evaporation of gold on a monolayer of spheres results in a sharp absorption spectrum, the presence of additional layers of spheres reduces the peak height as well as significantly broadening the half-width of the spectrum.

Figure 3 is a transmission electron microscope image of particles dislodged from the substrate; the sphere size is 110 nm and the deposition thickness is 20 nm. It clearly shows that only the top half of the particle becomes covered by gold.

1.4 Physical Parameters Under Control

We list below physical parameters that can be readily and reproducibly controlled.

Sphere Diameter

A large variety of monodisperse spheres with different diameters are available from many vendors. The diameter dictates the size of the cap-shaped particle as well as the distance between the particle and the underlying substrate. The surface quality of these spheres is also quite excellent.

Adsorption Density

While the dense coverage of the surface with a monolayer of spheres is over within ten minutes, by adjusting the incubation time one can control the adsorption density of surface-bound spheres; Fig. 4 shows SEM pictures of samples prepared with reduced incubation times. As the adsorption density is reduced, spheres tend to form independent clusters.

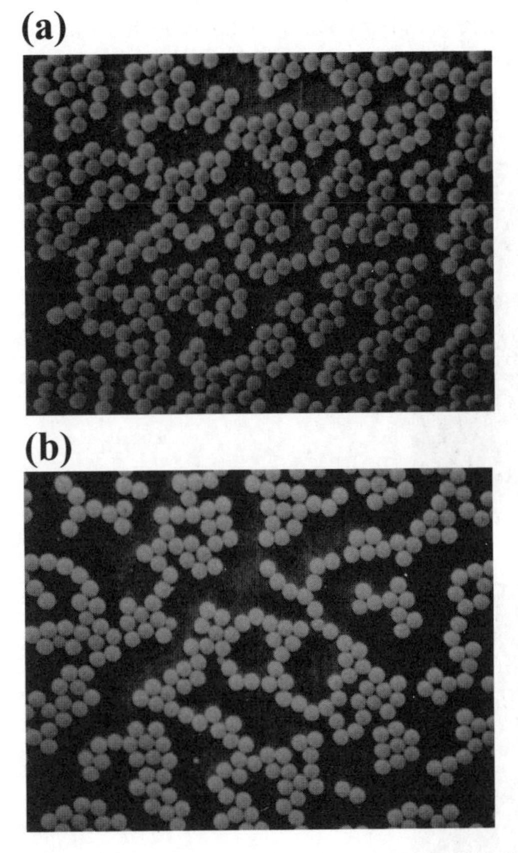

Fig. 4. SEM micrograph of adsorbed spheres (sphere diameter 209 nm) at reduced densities. Typical density at saturation is approximately 15.3 μm^{-2}, but by reducing the sphere incubation period, the density can be reduced. (**a**) Density of 12.2 μm^{-2}. (**b**) Density of 10.3 μm^{-2}

Deposition Thickness

One can readily control the deposition thickness of the metal layers. There are two layers involved, the one deposited directly on the substrate, and the other layer deposited on top of the sphere layer. Varying the thickness of each layer affects the optical property of the sample, but it is the layer deposited on top of the sphere monolayer that has the most profound effect.

Mixing Spheres of Different Diameters

While the strength of our method lies in its ability to produce monodisperse particles, it is possible to mix more than one kind of monodisperse spheres to obtain interesting results. Figure 5 shows such samples. The mixture consists of 110-nm spheres and 209-nm spheres, and two samples have been prepared from solutions of different mixture ratios.

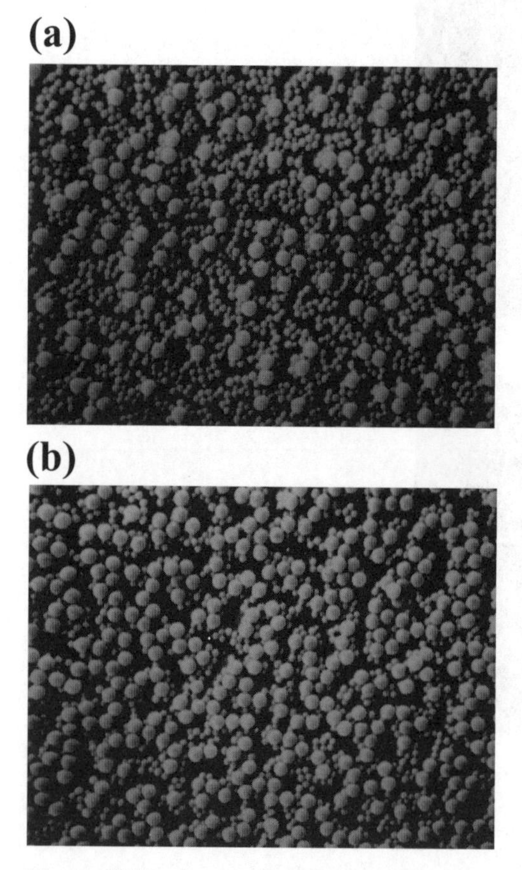

Fig. 5. SEM micrographs of samples prepared from a solution containing spheres of two different diameters, 209 nm and 110 nm. Two solutions with different mixture ratios were prepared. The resulting samples have adsorbed spheres with different densities (**a**) 110-nm spheres at 44.7 μm^{-2} and 209-nm spheres at 5 μm^{-2}. (**b**) 110-nm spheres at 20.4 μm^{-2} and 209-nm spheres at 10.1 μm^{-2}

Gap Control

Our method allows controlled formation of a gap between adjacent particles. As is well known from various studies on interactions among particles, the gap plays an important role on the optical properties. We can form a controlled gap by slight etching of PS spheres prior to deposition of gold on top of the sphere. The etching of surface-adsorbed PS spheres was first reported by Haginoya et al. [16] though their interest was primarily use of the etched PS spheres as a mask for subsequent evaporation of magnetic materials. As with Haginoya et al., we have also etched PS spheres by reactive ion etching (RIE) by using oxygen. The etching duration was adjusted to control the gap size. Figure 6 shows scanning electron micrographs of samples etched for different durations.

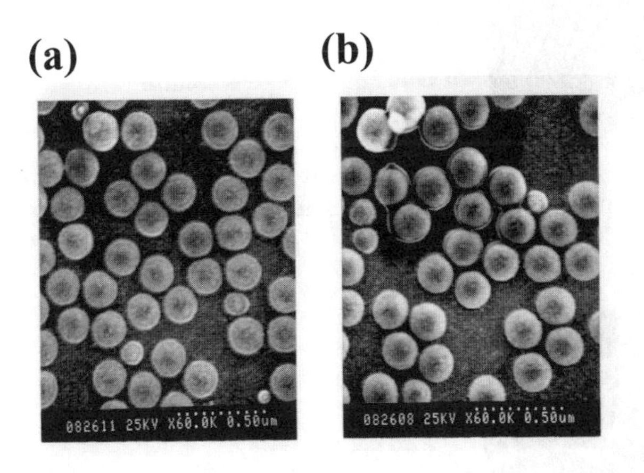

Fig. 6. SEM micrograph of spheres etched in oxygen plasma. The sphere size is 209 nm. (**a**) Etching duration of 60 s. (**b**) Etching duration of 50 s

Heat Treatment

To a lesser degree, we can control the shape with temperature. By raising the temperature slightly above the onset of polymer deformation, it is possible to deform the particle. Figure 7 is an SEM picture of a sample that has been heated up to 120°C for a brief period. The sphere has become deformed and now shows a greater contact area to the substrate. Interestingly, the presence of the gold cap has the effect of retaining the original shape of the sphere. In contrast, when surface-adsorbed PS spheres without gold are heated under the same condition, spheres simply turn into shapeless globs, as shown in Fig. 8.

Fig. 7. SEM micrograph of a sample temporarily heated up to 120°C. 209-nm spheres are covered by 20 nm of gold. The bottom half of the sphere becomes deformed, resulting in an increased contact area with the substrate. Image was taken at an angle of 45° from the surface normal

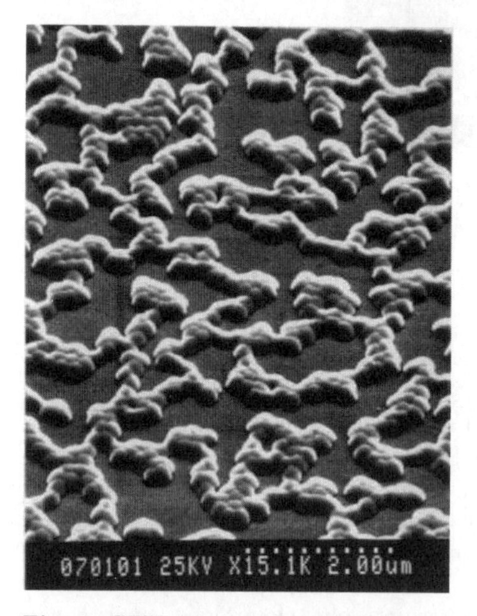

Fig. 8. SEM micrograph showing the effect of heating bare polystyrene spheres without gold on top. As with Fig. 7, the temperature was raised temporarily up to 120°C; in contrast to Fig. 7 lack of gold cap leads to complete merging of adjacent particles. Image was taken at an angle of 45° from the surface normal

Patterned Particles

We have found that we can control the adsorption process of spheres with patterns of alkanethiol molecules [11]. We prepared a gold-coated substrate consisting of regions of different alkanethiol molecules. By controlling the functionality of the alkanethiol molecule, it was possible to form patterned layers of spheres. To gain control on the adsorption process a distinct interaction between the spheres in suspension and selected areas of the surface needs to be introduced. While it is possible to utilize physical phenomena, such as changes in polarity, surface energy, or charge distribution in different regions of the surface, we have obtained the best results for chemically patterned surfaces in connection with the use of 1-ethyl-3-(3-(dimethylamino)propyl) carbodiimide (EDC). Here, the role of EDC in the adsorption process of the spheres is threefold. Due to its amino group, which is protonated in a neutral pH environment, the molecule is positively charged and thus behaves as a positively charged surfactant: it adsorbs on the negatively charged sphere surface, thereby changing the total charge of the spheres, i.e., their ζ-potential, and thus reducing interparticle repulsion. At the same time, the spheres are converted into carbodiimide functionalized particles and thus provide potential reaction sites for chemical bonding. Furthermore, dissolved excess EDC molecules act as macromolecules in the colloidal suspension and force aggregation due to depletion forces. We have found that we have to adjust the EDC concentration in suspension in such a way, that the body of EDC molecules adsorbs on the spheres and the remaining dissolved molecules are too few to exert depletion forces beyond kT. In this regime, which corresponds to 0.1–0.6 mM EDC (for suspensions with 2.6 vol.% spheres and particle sizes of some 100 nm) adsorption turned out to be sensitive to the chemistry of the surface. Note that this concentration is significantly lower than the value for preparing uniform, non-patterned samples. At higher EDC concentration, this sensitivity is screened by physisorption of the spheres due to depletion forces. We have proved this concept by replacing EDC with a poly(ethylene glycol) (PEG) molecule of similar size and molecular weight. PEG is non-adsorbing and chemically inert, so that it solely exerts depletion forces on the suspended spheres. At high concentration (above 0.6 mM), where the depletion forces exceed the Brownian motion of the suspended particles, we obtained similar results to those found for EDC at similar concentration. At low concentration, sphere adsorption failed independent of the chemical functionalities present at the surface. However, by using EDC and proper surface functionalization, e.g., by means of carboxyl-terminated alkanethiols, the carbodiimide functionalized spheres adsorb at low concentration due to chemical bonding. We have combined this selected chemical reactivity with the technique of microcontact printing to form patterned sphere layers. For that, a nonfunctionalized thiol (e.g., octadecanethiol; C_{18}) is printed onto a native gold surface by means of a microstructured elastomer stamp (Fig. 9). In this way, the C_{18} adsorbs only on the gold areas in contact with the stamp.

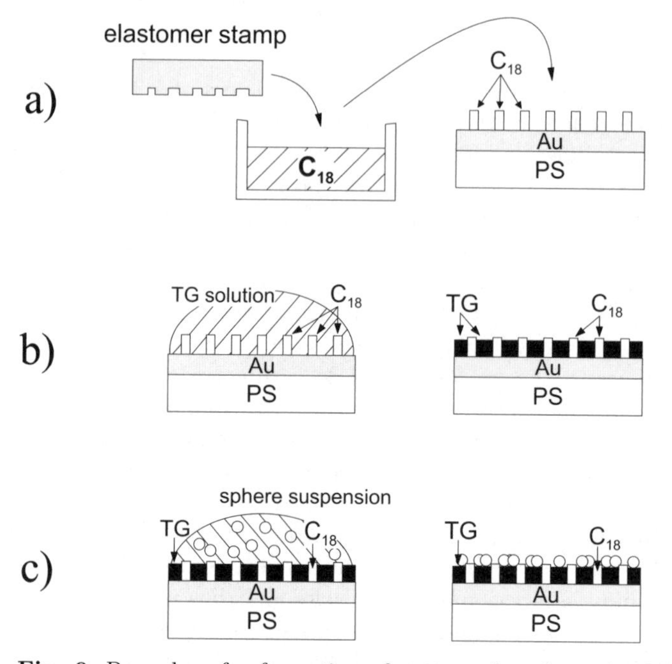

Fig. 9. Procedure for formation of patterned surface-adsorbed particles. (**a**) Immersion of a PDMS (poly dimethyl siloxane) stamp into an alkanethiol solution (octadecanethiol; C_{18}), and transfer of the thiol molecules onto a gold-coated polymer substrate. (**b**) Exposure of a second alkanethiol solution (thioglycolate; TG) to the substrate results in a pattern consisting of two types of alkanethiol molecules. (**c**) Exposure to a sphere suspension, and patterned sphere adsorption

In a second step, the gold substrate is immersed into a solution containing the functionalized alkanethiol molecule (e.g., carboxyl-terminated such as thioglycolate; TG), filling the remaining native areas on the surface. In this way, a chemical pattern of functionalized and nonfunctionalized areas has formed. Exposure of this pattern to an EDC-activated sphere suspension at low EDC concentration yields the formation of patterned sphere layers. Figure 10 is a SEM picture of a patterned sample.

1.5 Optical Properties

Figure 11 is a visible illustration of gold particles formed with our method. The entire region consists of a gold-coated (20 nm thick) PS substrate covered by a monolayer of 110-nm PS spheres; only selected regions were subjected to additional evaporation of 20 nm gold. The latter regions appear as dark letters due to strong absorption/scattering by cap-shaped gold particles; this

Fig. 10. SEM micrograph of patterned spheres. Spheres are selectively adsorbed on regions covered by thioglycolate separated by regions covered by octadecanethiol

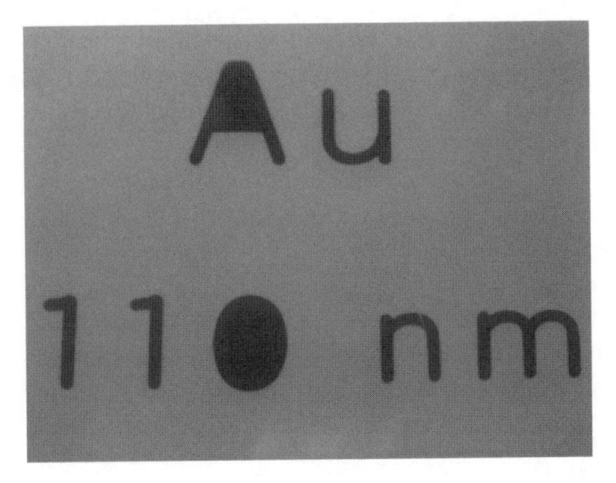

Fig. 11. Photograph of gold particles in the shape of letters. Twenty nanometers of gold were evaporated onto a polystyrene substrate, followed by formation of a uniform monolayer of spheres (110 nm in diameter). By using a stencil in the shape of letters, an additional 20 nm of gold were evaporated within well-defined regions. This is to illustrate visually the optical contrast formed by the gold particle

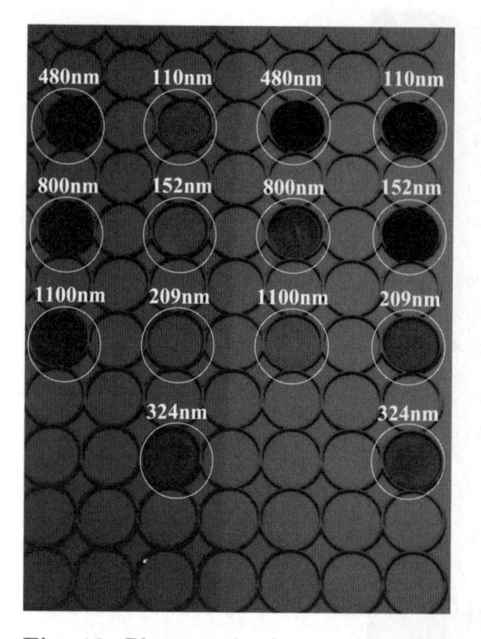

Fig. 12. Photograph of gold particles formed with PS spheres of different diameters. Spheres are adsorbed within circles defined by the rim while the number superimposed indicates the sphere diameter in nanometers. The spheres on the left-hand half of the sample are left bare whereas twenty nanometers of gold were evaporated on top of the spheres on the right-hand half

was accomplished by using a stencil with openings in the shape of letters. Although the monochromatic picture fails to convey the actual color, it is a striking violet. In Fig. 12, spheres of different diameters were subjected to the same treatment. Two identical sets of PS sphere monolayers, with the diameters being 110, 152, 209, 324, 480, 800 and 1100 nm, were prepared on the right and left halves of the substrate. Those on the left were left bare, and those on the right were subjected to evaporation of 20 nm gold. Although the monochromatic picture does not do it justice, in general, monolayers of larger PS spheres exhibit slight absorption without gold coating on top while not showing enhanced absorption after evaporation. With smaller spheres, diameters under a few hundred nm, formation of cap-shaped gold particles leads to significant enhancement in absorption.

Absorption Spectra in the Reflection Mode

The sphere diameter has a rather profound effect on optical properties. We have varied the sphere diameter between 55 nm and 1.1 μm but it is with spheres of smaller sizes, under a few hundred nm in diameter, where deposition of the top layer results in formation of markedly colorful particles.

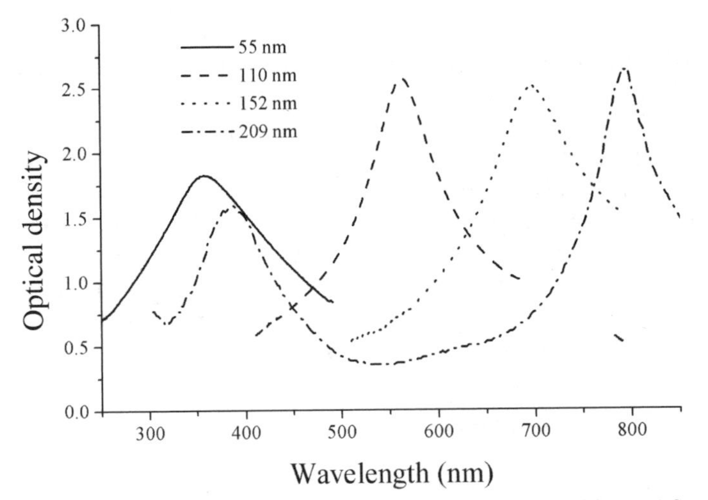

Fig. 13. Absorption spectra of surface-adsorbed gold particles in the reflection mode. Twenty nanometers of gold were evaporated onto spheres of different diameters. The reference is a 40-nm thick gold film.

Figure 13 shows absorption spectra of smaller PS particles in the reflection mode. While contributions from scattering and absorption have not been determined precisely, measurement of these samples within an integrating sphere indicates that absorption accounts for well over half of the effect. In contrast to the reflection mode, when absorption spectra of the same set of samples were measured in the transmission mode, there is hardly any absorption peak as shown in Fig. 14. For example, for 110-nm spheres, with a

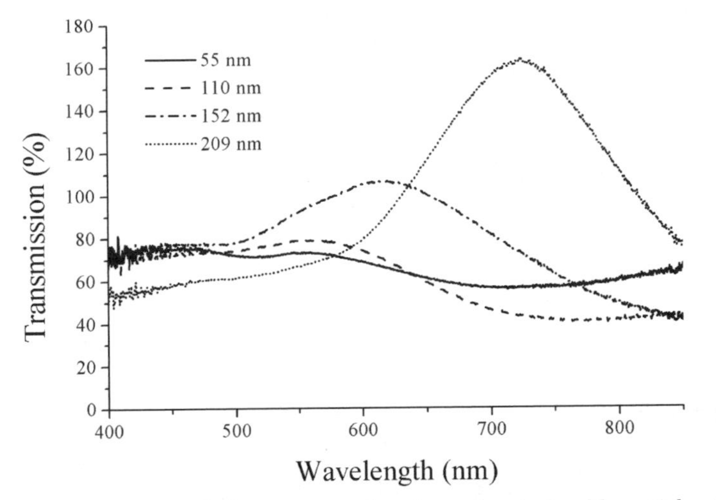

Fig. 14. Transmission spectra of surface-adsorbed gold particles. The reference is a 40-nm thick gold film; all samples have a 20-nm thick layer on the substrate and a 20-nm thick top layer on top of the sphere

prominent absorption peak around 550 nm in the reflection mode, there is no trace of reduced transmission around the same wavelength. Moreover, the presence of the PS particle layer can enhance transmission in some cases over a solid film of an equivalent total thickness; if a sample is prepared with both bottom and top gold layers being 20 nm thick, it can produce a higher transmission spectrum than a single 40 nm film even though the former exhibits a pronounced absorption spectrum in the reflection mode. The sample prepared with 209-nm spheres has a 65% enhancement in transmission around 720 nm. Even at 790 nm where it has the absorption peak in the reflection mode, the transmission is approximately 100 %. Thus, effectively, absorption by spheres more or less matches reflection by a plain gold film.

Deposition Thickness Dependence

The deposition thickness has been found to have a completely different effect depending on whether it is the top or bottom layer that is affected. While the bottom layer must be present in order to enhance the absorption spectrum, its exact thickness matters relatively little. On the other hand, the amount of deposited gold on top of the PS sphere has a great bearing on the absorption spectrum, as shown in Fig. 15 for 110-nm spheres. For other sphere diameters as well, initially a broad peak begins to form at longer wavelengths, and as the deposition thickness is increased, the peak height increases and shifts toward shorter wavelengths. Eventually, further increases fail to produce additional shifts. The saturation limit begins to set in at around 20 nm.

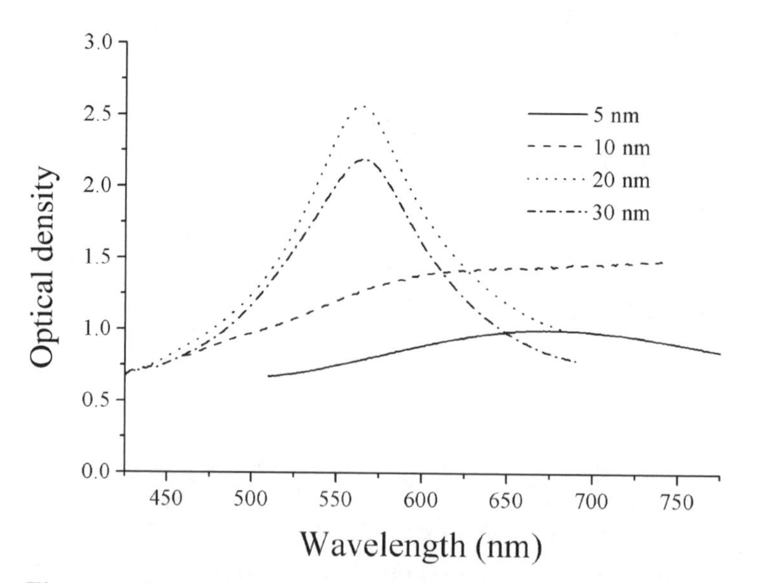

Fig. 15. Effect of the deposition thickness on the absorption spectrum. The sphere diameter is 110 nm for all the samples

Sphere Density

Figure 16 shows spectra of samples with different sphere densities. The density corresponds to 10.3 μm^{-2}, 12.2 μm^{-2}, and 15.3 μm^{-2} from sparse, moderate and dense samples, respectively, for 209-nm spheres in (a), and

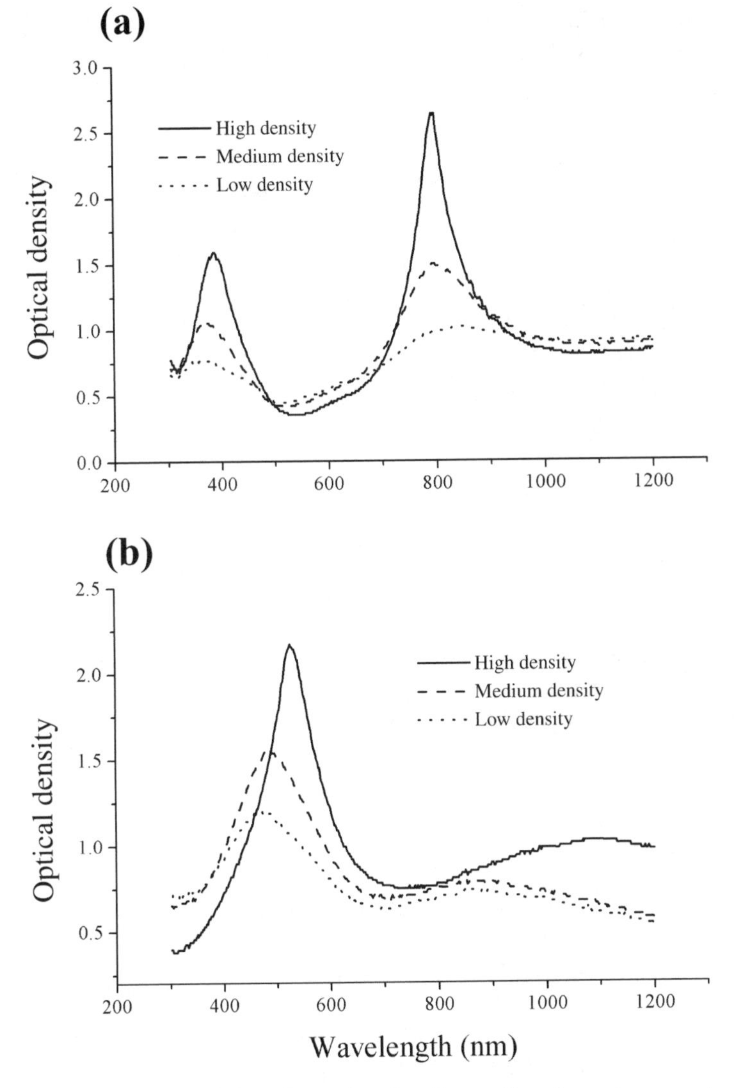

Fig. 16. Effect of the adsorption density on the absorption spectrum. (**a**) 209-nm spheres: the density is 10.3 μm^{-2}, 12.2 μm^{-2}, and 15.3 μm^{-2} from sparse, moderate and dense samples, respectively. (**b**) 110-nm spheres: the density is 39.8 μm^{-2}, 41.9 μm^{-2}, and 61.9 μm^{-2} from sparse, moderate and dense samples, respectively

39.8 μm^{-2}, 41.9 μm^{-2}, and 61.9 μm^{-2} from sparse, moderate and dense samples, respectively, for 110-nm spheres in (b).

Mixed Samples

Figure 17 shows spectra of mixed samples. Mixing two spheres of different diameters does not result in formation of two original peaks. Rather a single peak appears whose peak wavelength depends on the mixture ratio.

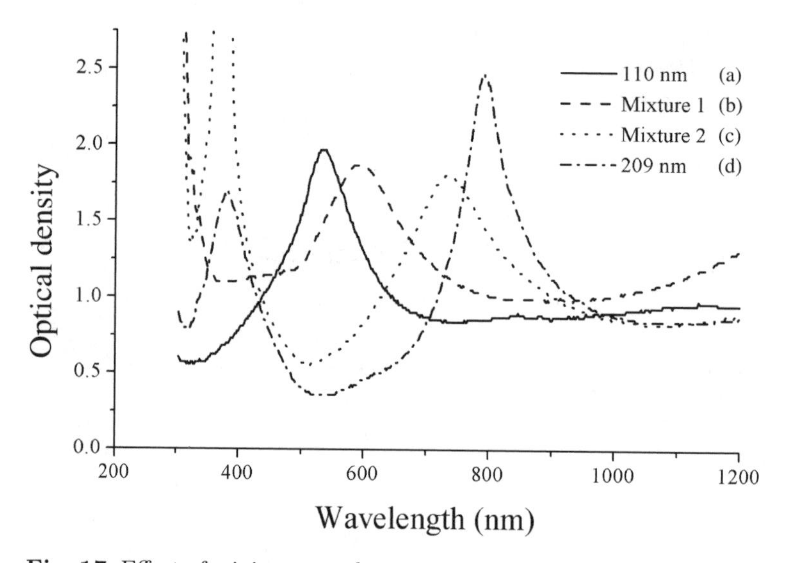

Fig. 17. Effect of mixing two sphere sizes at different mixture ratios. (**a**) 60.9 μm^{-2} for 110 nm. (**b**) 44.7 μm^{-2} for 110 nm and 5 μm^{-2} for 209 nm. (**c**) 20.4 μm^{-2} for 110 nm and 10.1 μm^{-2} for 209 nm. (**d**) 15.5 μm^{-2} for 209 nm

Material Dependence

We have concentrated our effort with gold particles simply due to its stability in air, but on occasion we have also experimented with other metals such as silver, copper and platinum. The spectra for silver samples are shown in Fig. 18. Twenty nanometers of silver were evaporated on spheres with diameters 55, 110, 152, 209 nm. They were quite susceptible to the atmosphere, as has been reported before in different systems [22].

Gap Dependence

The ability to control the gap distance at will between adjacent particles is one of the unique features of the current particle system. It is precisely this

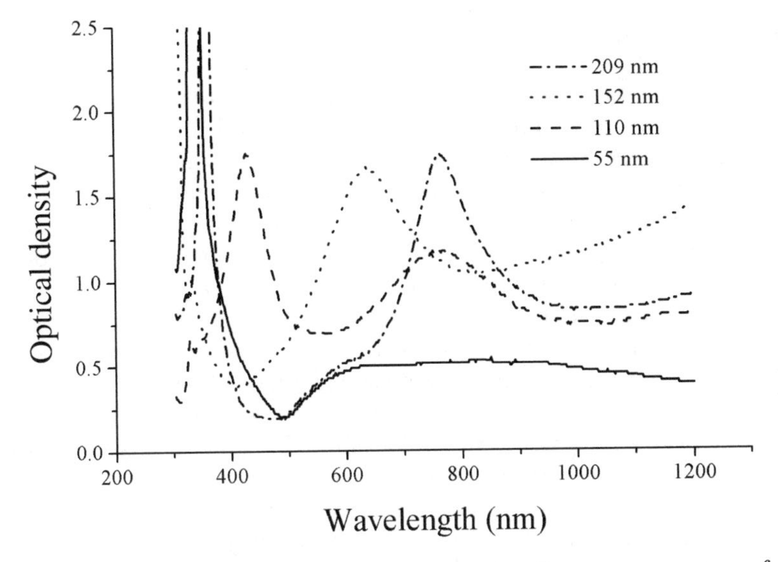

Fig. 18. Absorption spectra of silver particles. Twenty nanometers of silver were deposited on spheres of various diameters adsorbed on a 20-nm thick gold film

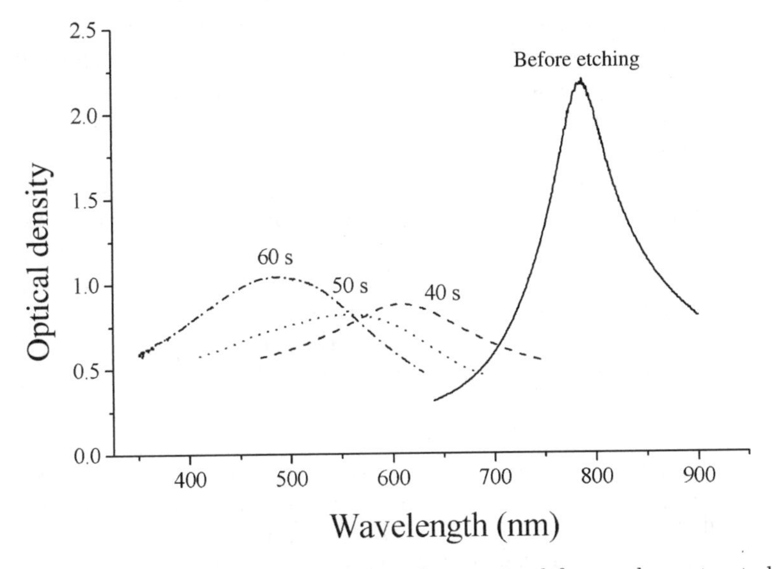

Fig. 19. Absorption spectra of samples prepared from spheres treated to oxygen plasma etching prior to deposition of the second gold layer. It can be seen that with increasingly longer etching periods, the absorption maximum shifts toward shorter wavelengths

kind of structure that is expected to affect the near-field effect in a profound way. Spectra of etched samples are shown in Fig. 19.

1.6 General Discussion

The present chapter deals mostly with a new methodology. We have presented some examples of what can be prepared and characterized them only in terms of absorption spectra and SEM micrographs. Most of the physical parameters that have been mentioned are well known to affect absorption spectra of noble metal particles; what is new here is the fact that these parameters can be, in any combination, modified reproducibly, and resulting samples are of sufficient size that makes macroscopic characterization an easy task. This is important not only for practical application of nano-optics based on metal particles, but also for rapid confirmation of theories with experimental data.

The size dependence is something that has always been part of the discussion over noble metal particles. While analysis is the most stringent for an isolated spherical solid particle of limited size, it has been extended to particles of different shapes such as ellipsoid, hemisphere, rod and cube [2]. In general, the fundamental mode of collective free-electron oscillation dominates in the quasistatic limit where the size is significantly smaller than the wavelength; modeling with free electrons applies well only for alkali metals, while noble metals, particularly gold, show pronounced deviations. For shapes other than a sphere such as the cap-shaped particle in this chapter, one would expect more than one fundamental mode even in the quasistatic limit. As it is, the sizes of particles prepared are beyond this limit, so that if the cap-shaped particle were suspended in isolation, one would expect to observe more than one peak. One might expect that if particles are oriented under an external force, it would be possible to selectively observe these peaks, but studies on cubic and hemispherical particles have revealed that the presence of edges and corners would preclude homogeneous polarization even when particles are oriented. Such an experiment is currently planned.

As the deposition thickness is increased, the diameter stays more or less constant as determined by the cross section of the polystyrene sphere, but the shape undergoes changes as the cap becomes thicker. In contrast to hemispherical particles and core-shell particles [2,20,23], cap-shaped particles have rarely been studied as far as we are aware. The core-shell particles, particularly with a dielectric core and a gold shell, have been receiving much interest in recent years [24–26], as they exhibit absorption peaks in the near infrared region, and the precise peak position can be controlled by the relative thickness of the shell. It can be imagined that as the deposition thickness is increased for the cap-shaped particle, the characteristics would shift from that of a core-shell particle to a hemispherical particle, if observed in isolation. Such an investigation is currently underway.

Difficulties of analyzing the actual samples prepared stem from the fact that particles are adsorbed on a substrate at high densities. Thus it would be necessary to take into account image interactions due to the substrate and interactions among adjacent particles. According to earlier studies on the effect of the substrate [23], multipole effects are enhanced and many secondary modes appear. In addition, with the sample under discussion here, the underlying gold film apparently constitutes part of an optical cavity, thus enhancing the overall spectrum [10,27].

The effect of interactions among adjacent particles is observed in more than one way. Varying the sphere density is one way, and the shift in the peak position as seen in Fig. 16 is one indication (the peak height is not a straightforward indicator of the sphere density as reduction in the density increases the area of pure gold surface that by reflection reduces the apparent peak height). Mixed samples are another indication. Particle systems with a certain size distribution are quite common, and they are characterized as mixtures of independent components. The spectra shown in Fig. 17 show, however, that spheres of two different sizes interact to give rise to a single new absorption peak. It should be mentioned that control over the sphere adsorption density has received attention recently [28].

Samples where a clear gap is formed among adjacent particles provide the clearest evidence for interactions. In recent years, there has been a revival of interest in coupled particles, particularly in connection with the enhanced near field in the gap region. There have been a number of theoretical reports [29,30], but experimentally it has proven difficult to create samples consisting of coupled particles with a well-controlled gap distance; it should be mentioned that Garcia-Santamaria et al. [31] have made an attempt with silica-coated gold nanoparticles. To our knowledge, we are reporting here for the first time an experimental technique that allows preparation of such samples. For now, we limit ourselves to the fact that the gap distance can be controlled by varying the etching duration, and the extent of the blue shift becomes greater with longer etching durations. There is no doubt that the etching also reduces the thickness of PS spheres so that this needs to be taken into account for future analysis. Moreover, to facilitate comparison with theoretical analysis, it would be desirable to prepare samples in such a way that the axis connecting all coupled particles would be the same because the relative angle between the axis and the polarization direction of the probing light plays an important role in the type of excited modes. One such sample would consist of parallel lines of stringed particles; it should be possible to prepare such samples by exploiting the patterning technique mentioned earlier.

It is quite exciting to speculate on and eventually verify experimentally the strength of the near field in the gap region. Theoretical work suggests [30] that the gap of coupled spheres is an ideal location to generate an intense "hot" spot that can be utilized for surface-enhanced Raman spectroscopy [32–34]. This might be verifiable by a combination of Raman spectroscopy and atomic force microscopy in the fashion reported by [35]; they attempted to correlate between the morphology of particle clusters and the strength of observed Raman signal but could not draw a conclusion on the shape of the ideal "hot" particle. Their conclusion, was however, that the intrinsic Raman enhancement factors of "hot" particle were on the order of 10^{14} to

10^{15} and only a very small number of particles contribute to enhancement. If we identify the characteristics necessary for enormous enhancement and selectively prepare them in a large number, we might be better able to exploit surface-enhanced Raman spectroscopy.

Besides surface-enhanced Raman spectroscopy, transmission of light through a film consisting of subwavelength structures has been receiving attention. In particular, an opaque metallic film consisting of a periodic array of subwavelength holes was found to transmit light of significantly longer wavelengths [36]. The transmission enhancement has been explained by a number of models including excitation of surface plasmon polaritons at the holes.

2 Optical Biosensing Application

Noble metal nanoparticles have long been known to exhibit an absorption spectrum that depends sensitively on the refractive index of the surrounding environment [2]. In recent years, as interest in life science has increased, many workers have proposed using this principle for nonlabeling detection of biomolecular interactions in a fashion similar to what has been accomplished with the surface plasmon resonance (SPR) sensor. Some have exploited changes in optical properties of metal particles upon aggregation induced by binding of biomolecules [37, 38], while others have made use of surface-adsorbed particles [5, 8, 18]. We will describe our biosensing technique [39, 40], starting with surface modifications with biomolecules and examples of actual measurement. Finally, we will emphasize one significant property of this novel biosensor; the "bulk" effect is rather diminished. The "bulk" effect refers to the fact although the SPR sensor is characterized as a surface-sensitive technique, it is actually influenced by bulk refractive index of the adjacent layer of fluid some few hundreds of nanometers in depth. Thus, when used as a biosensor, the sensor is sensitive toward both adsorption of molecules on the surface and presence of unbound molecules in the adjacent layer as well as fluctuations in salt concentration, making it not so trivial to separate these two effects. A diminished "bulk" effect of the current sensor, however, makes the novel biosensor rather robust against fluctuations in temperature and salt concentrations of the buffer used, while exhibiting a comparable sensitivity toward molecular binding.

2.1 Significance of Molecular Interactions in Life Science

There are many opportunities, ranging from medical applications in diagnostics and drug development to research applications in genomics and proteomics. Life, from the viewpoint of molecular biology, is fundamentally a manifestation of a myriad of biomolecular interactions in a wonderfully well-orchestrated fashion. Binding of a low molecular weight compound to a receptor molecule on a cell surface may trigger a series of reactions within the cell while binding of a promotor molecule to a DNA fragment initiates the expression of a gene. As the complete sequencing of the human genome comes to fruition, there will be even greater needs to screen binding events involving the products of these genes, in other words, proteins. With this in mind, we have selected biosensing as our principle target of application.

Experimental Setup

Figure 20 shows the sensor system consisting of a gold particle sensor, an optical fiber set up for illuminating the sensor with white light and collecting reflected light to be sent to a spectrometer, a custom software for peak-fitting and monitoring the peak position of the spectrum in real time. The gold particles were formed at the bottom of wells, 10 μl in volume, of the miniplates purchased from Nunc. The tungsten halogen lamp, LS-1, and the spectrometer, S2000, were both purchased from Ocean Optics. The fiber bundle equipped with SAM 905 connectors was fabricated by Tsuchida Seisakusho (Tokyo, Japan). The custom software is capable of fitting the spectrum with a pseudo-Voigt function in real time and displays the peak wavelength as a function of time. Four independent channels are provided so that four samples can be measured independently of each other. The setup is provided with a mixing mechanism to ensure that proper kinetic data can be obtained. The measurement protocol consists of the following steps. Figure 21 is a photograph of the setup.

1. The bottom of each well is coated with a capture biomolecule such as an antibody, a receptor protein, or a DNA primer.
2. A sample potentially containing the antigen is injected into the well.

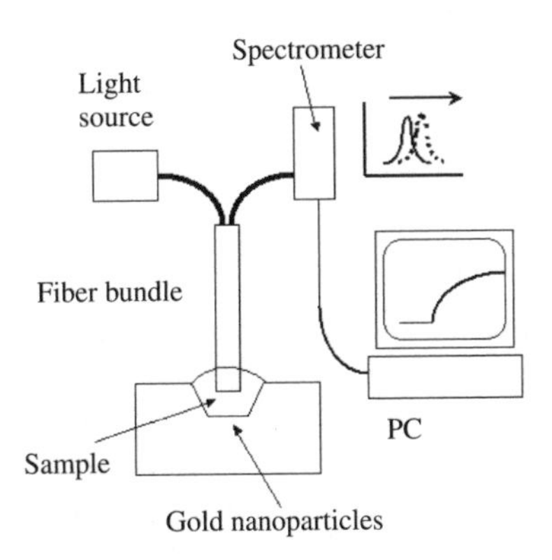

Fig. 20. Schematic diagram of the experimental setup. Gold particles were formed at the bottom of a polystyrene well with a volume of 10 μl. Light from a tungsten halogen lamp is led to the well bottom via a fiber bundle, and the reflected light is collected for real-time analysis by a spectrometer. The sample is mechanically agitated to assure immediate and uniform mixing

Fig. 21. Photograph of a four-channel model. A personal computer, not shown, completes the system

3. The reflectance spectrum is fitted by a pseudo-Voigt function in real-time, and the peak wavelength is displayed as a function of time.
4. The time change of the peak wavelength is fitted to an exponential function, and the time constant is obtained.
5. The time constant is obtained from samples with different concentrations, and the affinity is calculated.

Surface Coating

There are many methods for adsorbing biomolecules onto the sensor surface.

(1) Direct physisorption

It is well known that many biomolecules bind irreversibly to a bare gold surface. It is thus possible to inject a buffer solution containing a biomolecule. The selection of the buffer solution is not crucial, but we often use phosphate buffer saline (PBS) maintained at pH 7.4. Although the concentration of 1 mg/ml quickly saturates the surface in a matter of a few minutes, the concentration of 100 μg/ml is enough with a slightly longer incubation period.

(2) Coupling via protein A or protein G

Direct physisorption described above is rather straightforward, but there is some concern that the antibody may partially lose its ability to bind its antigen through denaturation. This possibility is reduced if the antibody is adsorbed indirectly via an intermediate layer. Protein A and protein G can serve such a purpose; they both have a multiple number of binding sites for antibodies (four for protein A and two for protein G) so that physisorbed protein A or protein G is likely to retain its function. As in (1), protein A or protein G at the concentration of 100 μg/ml can be injected into the sensor well, and after some 10 minutes, the coverage is more or less complete. Then, a solution containing an antibody is injected. The origin of the antibody, whether from mouse, goat, human, etc., has some effect on the affinity of protein A or protein G toward the antibody.

(3) Coupling via biotin–avidin (streptavidin) reaction
The great affinity between biotin and avidin (or similarly for streptavidin) is well known, and most biomolecules can be readily modified by a biotin moiety; thus physisorbed avidin can serve as the binding layer for a biotinylated biomolecule.

(4) Activated carboxyl group
Formation of a covalent bond between an activated carboxyl group and an amino group is widely used as a linking method; most biomolecules possess an amino group so that a surface-bound carboxyl group can serve as a starting point. Gold is a convenient surface because an alkanethiol molecule is known to readily form a self-assembled monolayer (SAM) and alkanethiol molecules with carboxyl functional groups are commercially available. An ethanol solution containing 10-carboxy-1-decane thiol (10-carbo.), purchased from Dojindo, is prepared at the concentration of 1 mM. Exposure of the sensor well to the solution for a period of 10 minutes is enough to prime the sensor surface with a carboxyl group. A freshly prepared mixture of EDC and N-hydroxy succinimide (NHS), 0.1 M, is added for some 10 minutes and immediately afterwards the biomolecule to be adsorbed is injected.

2.2 Measurement Examples

Antigen–Antibody Reaction 1

One of the most widespread uses of a biosensor is to detect an antigen with a corresponding antibody. Here we show in Fig. 22 a result of detecting a low molecular weight peptide (molecular weight of 841 dalton), a polypeptide of 6 histidine (6 His) amino acid residues, with an anti-6 His antibody. First, streptavidin protein is directly physisorbed onto the bare gold surface. Then, biotinylated protein A is coupled to streptavidin via biotin/streptavidin binding. Finally, the anti-6 His antibody is anchored to protein A via the F_c region. Note that each time a molecule binds to the sensor surface, the peak wavelength increases. Once the sensor is coated with an antibody, it is ready to detect an antigen. When 6 His is added to the sensor at the concentration of 0.5 mg/ml, despite its low molecular weight, there is an observable increase

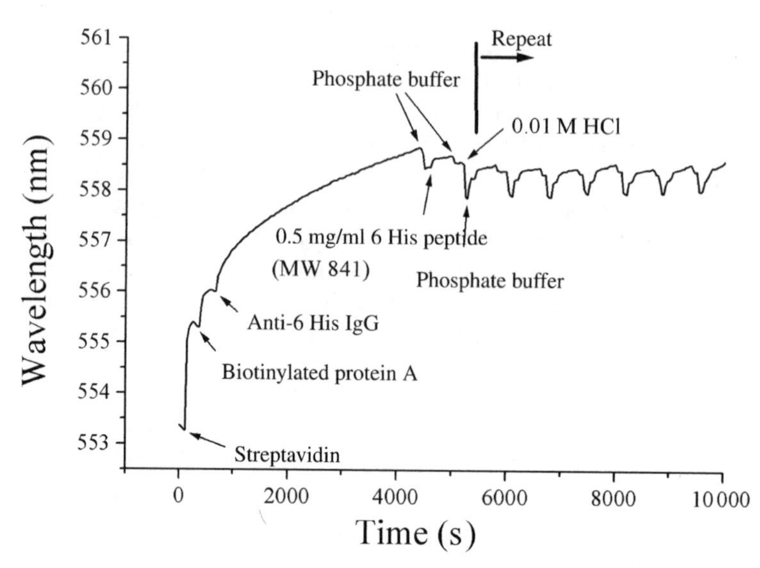

Fig. 22. Antigen–antibody reaction using a polypeptide (6 His: a string of six histidine amino acid residues) as the antigen. In order to bind an anti-6 His antibody (IgG: immunoglobulin G), the sensor surface is exposed to streptavidin, biotinylated protein A and anti-6 His antibody sequentially. When captured by protein A, the antibody exposes its binding sites toward the liquid phase, away from the substrate. The surface-adsorbed antibody can be used a number of times; a capture antigen molecule, 6 His in this example, can be released by exposure to diluted 10 mM hydrochloric acid, and the antibody is ready to capture another antigen molecule again. Here, the sequence is repeated 8 times

in the peak wavelength. Once the binding becomes saturated, the unbound antigen is washed off with phosphate buffer. To remove the antigen, addition of diluted hydrochloric acid, 10 mM, is enough to break up the antigen–antibody reaction. This is reflected in a reduction in the peak wavelength. Addition of diluted acid, however, leaves the antibody intact so that when the antigen is added again, the surface-bound antibody can capture it. This process can be repeated.

Antigen–Antibody Reaction 2

Here we show another example of an antigen–antibody reaction. This time streptavidin protein is directly physisorbed onto the sensor as in the previous example. However, surface-bound streptavidin is used to capture an anti-streptavidin antibody. As the molecular weight of an antibody (IgG: immunoglobulin G) is rather high at 150 kilodalton, it can be detected at significantly reduced concentrations. Samples of different concentrations were added to the sensor at time $t = 200$ s, and all the results were superimposed. After 600 s, the sensor was rinsed with phosphate buffer saline, followed by

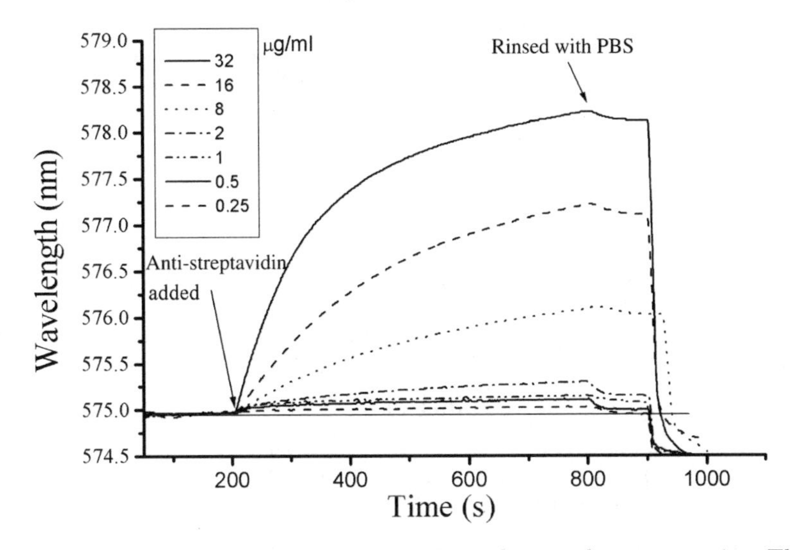

Fig. 23. Dependence of the sensorgram on the sample concentration. The covered range is from 0.25 μg/ml to 32 μg/ml. At t = 200 s, anti-streptavidin antibody is injected onto a sensor covered by streptavidin. At t = 800 s, phosphate buffer saline (PBS) is used to rinse off unbound antibody. After t = 900 s, the sensor is exposed to diluted hydrochloric acid to break the bond between streptavidin and anti-streptavidin

injection of diluted hydrochloric acid to remove the antibody completely. This was carried out for samples of different concentrations as shown in Fig. 23.

The purpose of injecting samples of different concentrations was to obtain kinetic data. Quite often the main goal of measuring binding reactions is to obtain the equilibrium binding constant k_D. This can be obtained by repeating the same reaction at different concentrations and each binding curve is fitted by an exponential function. The time constant is plotted against the concentration. The slope is equivalent to k_{ass} and the y intercept is equivalent to k_{diss}. k_D is simply k_{diss} divided by k_{ass}. Figure 24 shows an example of such a plot. The quality of fitted data is comparable to that obtained with a commercial instrument with significantly increased complexity.

Protein–DNA Interactions

Figure 25 shows a binding event between a surface-adsorbed double-stranded DNA and a DNA-binding protein.

2.3 General Characteristics of the Sensor

While there are many types of biosensors [6, 8, 41–44], one of the most important parameters for a biosensor is sensitivity. Naturally it is roughly de-

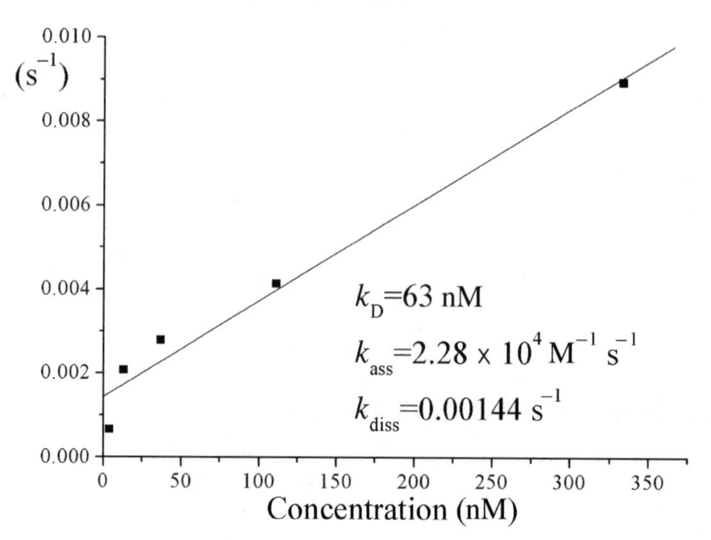

Fig. 24. Obtaining kinetic data. Sensorgrams from Fig. 23 are fitted to a single exponential function, $A - B\exp(-kt)$, and the binding constant, k, was obtained for various concentrations. All the binding constants were plotted as a function of the concentration; linear regression was used for fitting. The slope of the fitted data is the association time constant, k_{ass}, and the y intercept is the dissociation time constant, k_{diss}. The equilibrium dissociation constant, k_{D}, is $k_{\mathrm{diss}}/k_{\mathrm{ass}}$. k_{D} of 63 nM is fairly typical for an antigen–antibody reaction. The quality of linear regression also indicates the reliability of our sensor

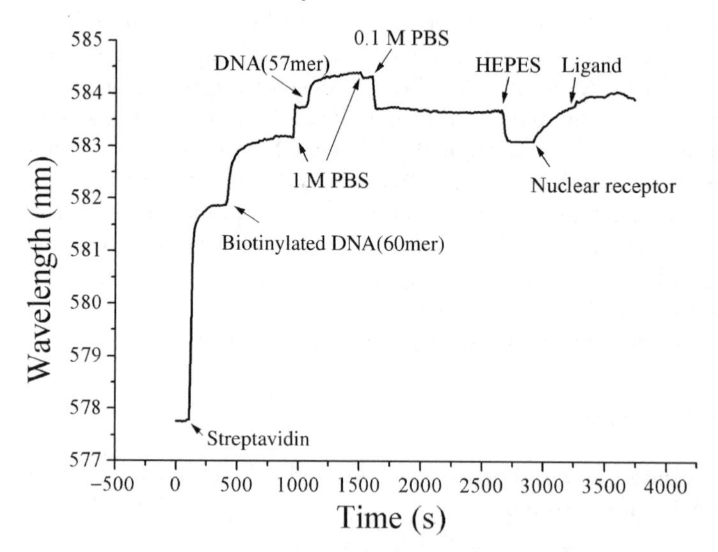

Fig. 25. Sensorgram showing DNA hybridization followed by binding of a DNA-binding protein. A biotinylated 60 mer was first attached via surface-bound streptavidin, and the second DNA strand, 57 mer, was hybridized in a high salt concentration buffer. After rinsing with HEPES buffer, a nuclear receptor, RXR, was injected

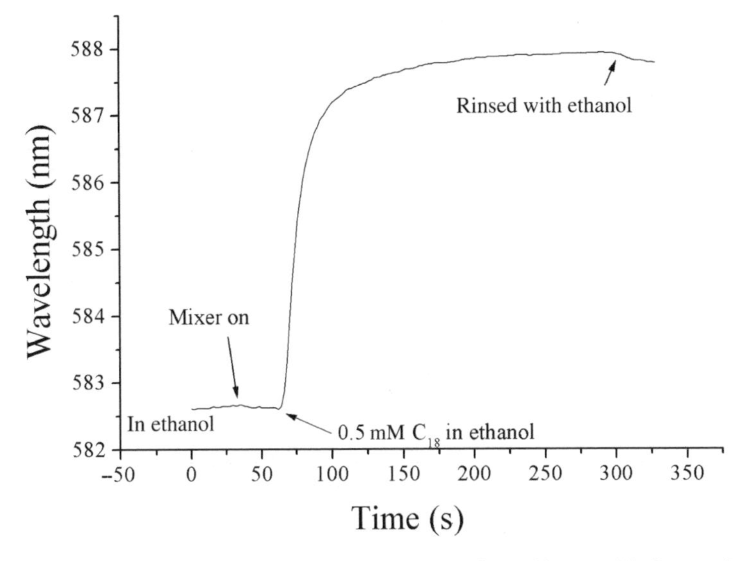

Fig. 26. Sensorgram showing formation of a self-assembled monolayer (SAM) of octadecanethiol. Because SAM has been well characterized, it is used here as a calibration standard

termined by the amount of shift induced when a molecular layer of certain density and thickness forms and the minimum shift that can be detected.

Figure 26 shows a sensorgram when a self-assembled monolayer (SAM) of octadecanethiol forms on the sensor with the full shift of approximately 5.5 nm in the peak wavelength. According to the literature [45, 46], the full thickness of such a SAM is 2 nm. The noise level obviously depends on many factors such as the specification of the spectrometer, intensity of the light source, throughput of the optical fiber system, as well as processing of the signal. Thus, what is shown in Fig. 27, stability of the baseline when no reaction is taking place, is only a rough indication of what the current system is capable of; if the shift in the peak wavelength is more than 0.02 nm, it can be detected without difficulty. Thus, it is safe to say that a submonolayer with an averaged thickness of 0.01 nm can be detected.

Refractive Index Dependence

The sensor is exposed to various fluids of different refractive indices. This was achieved by preparing mixtures of water and glycerol with different mixture ratios as shown in Fig. 28. With refractive indices of 1.33 for water and 1.47 for glycerol, the sensorgram shows that the change in the peak wavelength for a unit change in the refractive index is 61 nm. In contrast, the literature figure for the traditional SPR, when measured in terms of wavelength, is typically 3000 nm per unit change in the refractive index [42, 47]. This

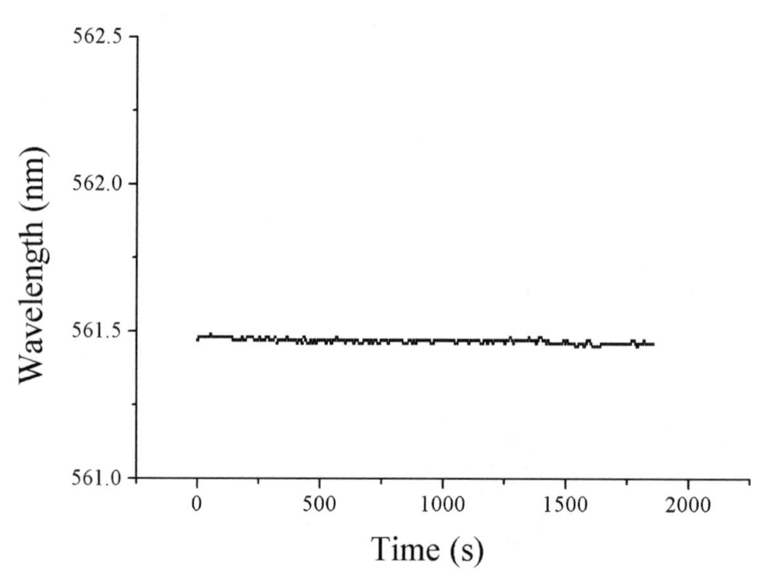

Fig. 27. Sensorgram showing stability of the baseline

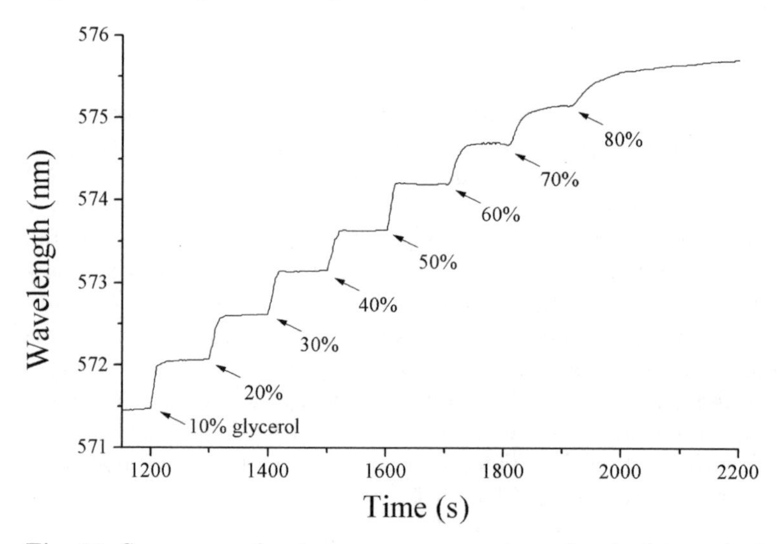

Fig. 28. Sensorgram showing sensor response toward water/glycerol mixtures. The volume concentration of glycerol is incrementally increased by 10%; the first mixture contains 10% glycerol, and the last mixture added at $t = 1900$ s has 80% glycerol. With refractive indices, $n = 1.33$ for water and 1.47 for glycerol, it is roughly estimated that the shift is 61 nm per one refractive index unit, $\delta n = 1$

would suggest that the sensitivity of our sensor is significantly lower. Figure 29, however, shows an interesting comparison between the SPR sensor and our sensor subjected to an identical protocol; the first half consists of a

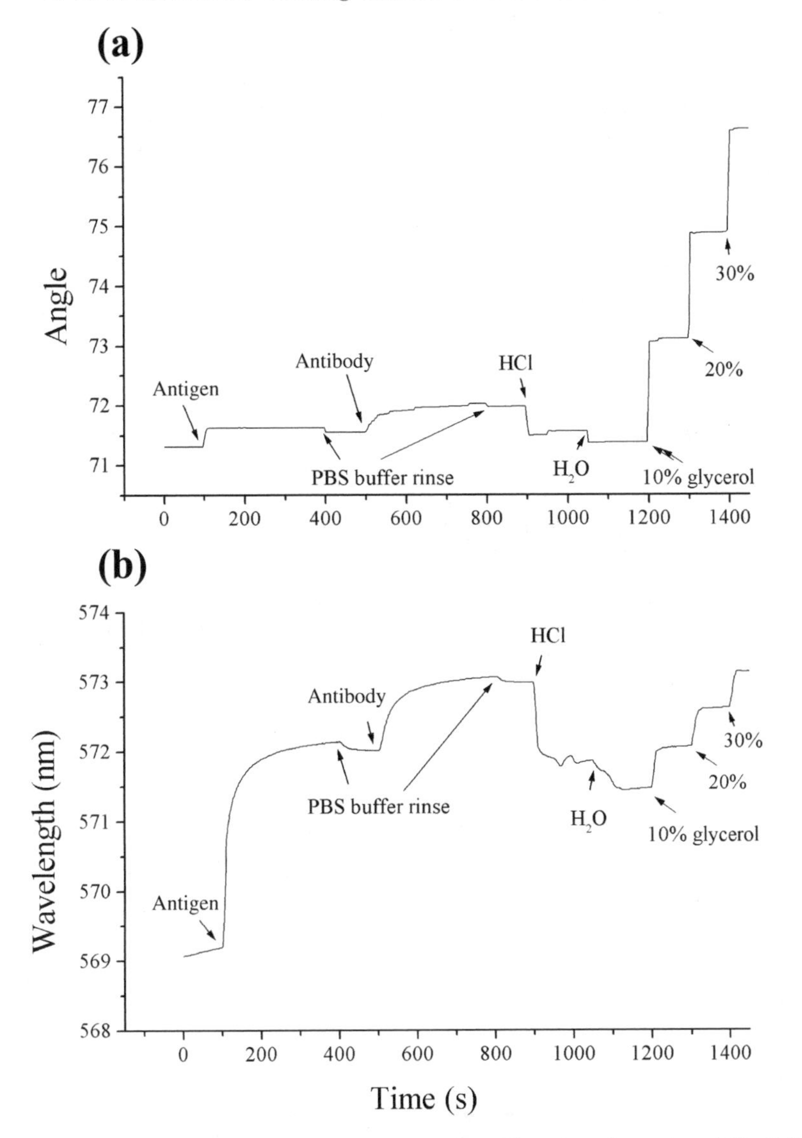

Fig. 29. The following protocol was used to illustrate the reduced "bulk" effect of our LSPR instrument with respect to the SPR instrument; at $t = 100\,\mathrm{s}$, 1 mg/ml avidin was added to the sensor to initiate physisorption. After saturation, unbound avidin was rinsed with PBS buffer at $t = 400$ s. 100 μg/ml of anti-avidin antibody was injected to start the antigen–antibody reaction; the overall shift in the wavelength is typical. The portion of the sensorgram after $t = 1200$ s is identical to the initial portion of Fig. 28. (**a**) Response of an SPR sensor from Texas Instruments, Spreeta, that was chosen as a representative SPR sensor. (**b**) Our instrument. Note how the shifts induced by injection of glycerol mixtures are suppressed in (**b**)

regular antigen–antibody reaction followed by an incremental increase in the refractive index of the fluid. At 100 s, avidin was physisorbed on the sensor, followed by injection of an anti-avidin antibody at 500 s. Then, starting at 1200 s, water/glycerol mixtures are injected every 100 s, with increasingly higher mixture ratios. Note that shifts in the latter half of the protocol, in comparison to the first half, are significantly greater with the SRP than with our sensor. This shows that our sensor has a significantly enhanced response toward binding of a biomolecule over that induced by change in the refractive index of the surrounding medium [48].

Temperature Dependence

Figure 30 shows the response of the sensor when water at various temperatures is injected. Initially it is at room temperature (20°C), and the temperature of water injected is increased incrementally from 1 to 63°C; each injection results in a peak because the temperature returns to room temperature rather rapidly. There are two noteworthy features. One is that even though the refractive index is higher for water at lower temperature in this temperature range, the shift in the peak wavelength is opposite; an increase in

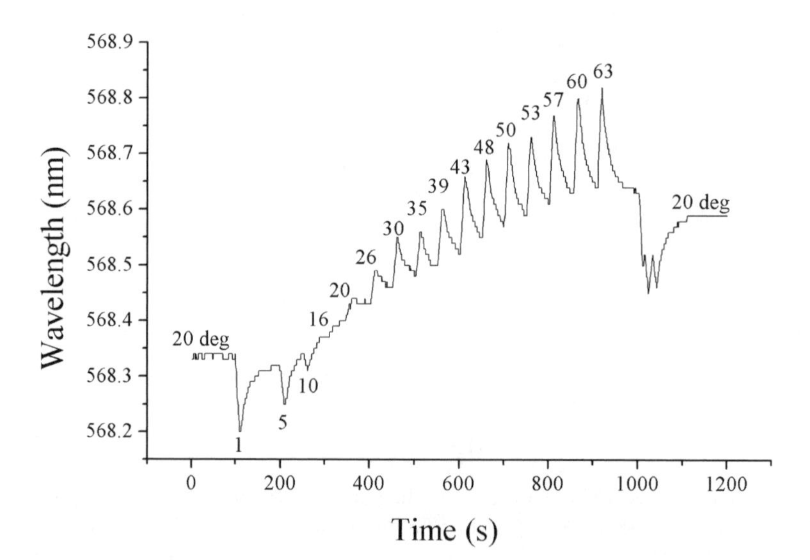

Fig. 30. Temperature dependence of the sensor. Water of various temperatures was injected. The experiment starts at room temperature, 20°C. Temperature of water injected was 1°C at $t = 100$ s, 5°C at $t = 200$ s, 10°C at $t = 250$ s, 16°C at $t = 300$ s, 20°C at $t = 350$ s, 26°C at $t = 400$ s, 30°C at $t = 450$ s, 35°C at $t = 500$ s, 39°C at $t = 550$ s, 43°C at $t = 600$ s, 48°C at $t = 650$ s, 50°C at $t = 700$ s, 53°C at $t = 750$ s, 57°C at $t = 800$ s, 60°C at $t = 850$ s, 63°C at $t = 900$ s The temperature returns to room temperature within tens of seconds. Between $t = 1000$ and 1050 s, water at 1°C was injected three times

the density results in a decrease in the peak wavelength. This suggests that a change in temperature has an effect on the sensor through a mechanism other than a simple change in the refractive index, possibly a structural change in the nanostructure of the sensor itself. The second noteworthy feature is that there is hysteresis; the peak wavelength at room temperature, after the series of temperature jumps, is permanently changed. Again, this might reflect a structural change in the nanostructure of the sensor itself. While this issue might call for attention in the future, the fact that temperature-induced shifts are quite small and most if not all binding experiments will not subject the sensor to such large jumps in temperature renders this a minor nuisance rather than a problem. In contrast, a typical SPR instrument is characterized by a jump of 0.01 angular degree (e.g., 100 resonance units in the industry parlance) per 1 degree Celsius [49]. In terms of the wavelength of our instrument, it would loosely translate to 0.5 nm/degree Celsius.

In summary, we have developed a novel biosensor based on optical properties of gold nanoparticles. There have been numerous reports in recent years on similar sensors also based on localized surface plasmon resonance, LSPR, of gold nanoparticles [6, 8]. Advantages cited for the LSPR sensor over the SPR sensor are simplicity of the instrument, low costs, ability to give a result that can be read off visually, etc. On the other hand, there have been few reports on the actual performance. It has been observed that the dependence of the peak wavelength on the refractive index of the surrounding material is significantly lower, by more than one order of magnitude. We also make similar observations, but we emphasize that as far as detection of biomolecules is concerned, it has a sensitivity significantly higher than what the result of the bulk dependence leads one to believe. The reduced "bulk" effect is a tremendous advantage because molecular-binding events can be monitored stably in the presence of fluctuations in temperature and salt concentrations of the buffer solution.

One outstanding characteristic of our LSPR sensor is the extraordinarily high optical density and the narrow width of the absorption peak. Typically the optical density of surface-adsorbed gold nanoparticles is less than 0.5, but the figure for our sensor is more than four times greater. This facilitates accurate determination of the peak wavelength, better than 0.02 nm. As the shift induced upon formation of a SAM, corresponding to a mass loading of 1 ng/mm^2, is typically 5.5 nm, under the assumption that linearity holds, this leads to the detection limit of under 5 pg/mm^2. This compares well with the reported practical limit of 10 pg/mm^2 for the SPR sensor.

3 Curious Observation

While in most cases, the response of our sensor is quite comparable to that of the SPR sensor with the generally observed difference mentioned earlier, some protocols have led to curious observations. One such case is shown in

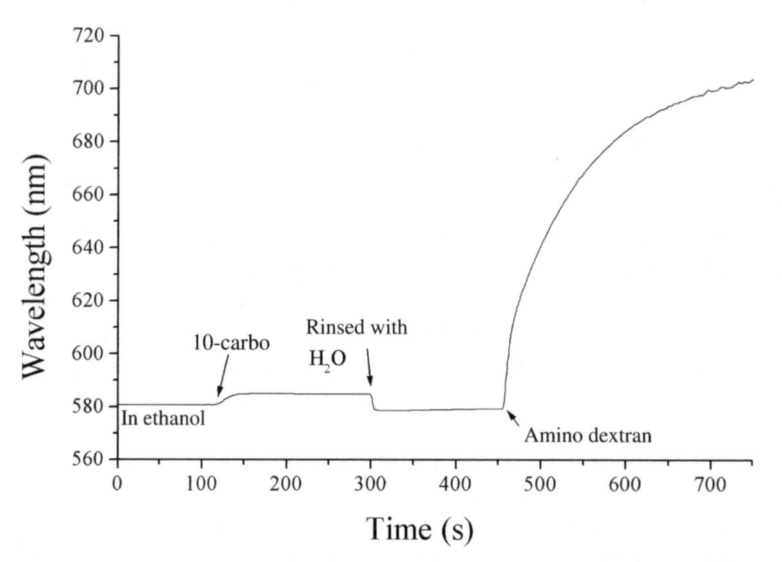

Fig. 31. Binding of amino dextran (MW 10 kilodalton) onto the sensor coated with 10-carbo. Ostensibly, the negative charge of the thiol molecule attracts electrostatically positively charged amino dextran suspended in water. Remarkably, the shift is well over 120 nm. If the refractive-index dependence in Fig. 28 is any indication, this would reflect change of more than two refractive index units. Obviously, this cannot be so, thus inviting further exploration for an explanation

Fig. 31. Here, The sensor was first exposed to an ethanol solution containing 10-carbo at the concentration of 1 mM. Then, excess 10-carbo was rinsed with pure water, resulting in reduction in the peak wavelength. Then, a solution containing 10 kilodalton amino dextran was injected. Opposite charges on the surface-bound carboxyl group and amino dextran are expected to lead to formation of an amino dextran layer. Indeed this was what we observed, but the shift was phenomenal at well over 120 nm. Although the exact shift depends on the molecular weight of the amino dextran used and its concentration, such phenomenal shifts are routinely observed with this particular protocol. When we recall that a shift due to a SAM formation was well under 10 nm, and a shift caused by a change in the refractive index was some 61 nm per unit change, a shift of more than 100 nm is surprising. While we suspected a permanent change in the sensor structure, observation of the sensor with a scanning electron microscope did not reveal any noticeable change, not to mention removal of the spheres.

4 Conclusion

We have shown a new method for forming surface-adsorbed metal particles; The beauty of this method is its simplicity and practicality. While this method will certainly give rise to many applications, we have concentrated on its application to biosensing where currently there is a great demand for the emerging field of proteomics. In comparison to existing analytical techniques, the LSPR sensor based on gold nanoparticles produced with the new method has turned out to possess a number of advantages. It is characterized by a significantly reduced "bulk" effect, which is a great plus as it allows stable read out in the presence of fluctuations in temperatures and salt concentrations.

Acknowledgements

This work has been made possible by a grant MF-16 from the Ministry of Health and Labor. We express much appreciation to Dr. T. Fujimura, Dr. J. Pipper, Dr. Z. Cao, Dr. T. Okamoto, Dr. H. Sugiyama, Dr. K. Ninomiya, Dr. T. Iwayanagi, and Dr. H. Taguchi in the course of this project. Without their help, the project would not have been possible. We also thank the Ministry of Health and Labor, and Dr. Inoue of the National Institute of Health Science of Japan for their partial support of this project. Professor M. Furuya and Dr. H. Kambara have been constant sources of encouragement.

References

1. T. Kume, N. Nakagawa, S. Hayashi, K. Yamamoto: Solid State Commun. **93**, 171 (1995)
2. *Optical Properties of Metal Clusters* by U. Kreibig, M. Vollmer (Springer, Berlin 1995)
3. *Nanoparticles and Nanostructured Films* ed. by J.H. Fendler (Wiley-VCH, Weimheim 1998)
4. R. Micheletto, H. Fukuda, M. Ohtsu: Langmuir **11**, 3333 (1995)
5. G. Kalyuzhny, M.A. Schneeweiss, A. Shanzer, A. Vaskevich, I. Rubinstein: J. Am. Chem. Soc. **123**, 3177 (2001)
6. F. Meriaudeau, T. Downey, A. Wig, A. Passian, M. Buncick, T.L. Ferrell: Sens. Actuators B **54**, 106 (1999)
7. M. Brust, D. Bethell, C.J. Kiely, D.J. Schiffrin: Langmuir **14**, 5425 (1998)
8. T. Okamoto, I. Yamaguchi, T. Kobayashi: Opt. Lett. **25**, 372 (2000)
9. G.A. Niklasson, H.G. Craighead: Thin Solid Films **125**, 165 (1985)
10. H. Takei: J. Vac. Sci. Technol. B **17**, 1906 (1999)
11. M. Himmelhaus, H. Takei: Phys. Chem. Chem. Phys. **4**, 496 (2002)
12. A.S. Dimitrov, C.D. Dushkin, H. Yoshimura, K. Nagayama: Langmuir **10**, 432 (1994)
13. T. Fujimura, T. Itoh, A. Imada, R. Shimada, T. Koda, N. Chiba, H. Muramatsu, H. Miyazaki, K. Ohtaka: J. Lumino. **954**, 87 (2000)
14. Y. Masuda, M. Itoh, T. Yonezawa, K. Koumoto: Langmuir **18**, 4155 (2002)
15. H. Takei, N. Shimizu: Langmuir **13**, 1865 (1997)
16. C. Haginoya, M. Ishibashi, K. Koike: Appl. Phys. Lett. **71**, 2934 (1997)

17. J.C. Riboh, A.J. Haes, A.D. McFarland, C.R. Yonzon, R.P. Van Duyne: J. Phys. Chem. B **107**, 1772 (2003)
18. M.D. Malinsky, K.L. Kelly, G.C. Schatz, R.P. Van Duyne: J. Am. Chem. Soc. **123**, 1471 (2001)
19. U.C. Fischer, D.W. Pohl: Phys. Rev. Lett. **62**, 458 (1989)
20. P. Barnickel, A. Wokaun: Mol. Phys. **67**, 1353 (1989)
21. G.L. Hornyak, C.R. Martin: Thin Solid Films **303**, 84 (1997)
22. Y.W. Cao, R. Jin, C.A. Mirkin: J. Am. Chem. Soc. **123**, 7961 (2001)
23. R.J. Warmack, S.L. Humphrey: Phys. Rev. B **34**, 2246 (1986)
24. H.S. Zhou, I. Homma, H. Komiyama, J.W. Haus: Phys. Rev. B **50**, 12052 (1994)
25. R.D. Averitt, D. Sarkar, N.J. Halas: Phys. Rev. Lett. **78**, 4217 (1997)
26. T. Pham, J.B. Jackson, N.J. Halas, T.R. Lee: Langmuir **18**, 4915 (2002)
27. A. Leitner, Z. Zhao, H. Brunner, F.R. Aussenegg, A. Wokaun: Appl. Opt. **32**, 102 (1993)
28. R.R. Bhat, D.A. Fischer, J. Genzer: Langmuir **18**, 5640 (2002)
29. J.P. Kottmann, O.J.F. Martin: Opt. Lett. **26**, 1096 (2001)
30. H. Xu, M. Käll: Phys. Rev. Lett. **89**, 246802 (2002)
31. F. Garcia-Santamaria, V. Salgeuirino-Maerira, C. Lopez, L.M. Liz-Marzan: Langmuir **18**, 4519 (2002)
32. P.F. Liao, M.B. Stern: Opt. Lett. **7**, 483 (1982)
33. R.G. Freeman, K.C. Grabar, K.J. Allison, R.M. Bright, J.A. Davis, A.P. Guthrie, M.B. Hommer, M.A. Jackson, P.C. Smith, D.G. Walter, M.J. Natan: Science **267** 1629 (1995)
34. K. Kneipp, Y. Wang, H. Kneipp, L.T. Perelman, I. Itzkan, R.R. Dasari, M.S. Feld: Phys. Rev. Lett. **78**, 1667 (1997)
35. S. Nie, S.R. Emory: Science **275**, 1102 (1997)
36. A. Dogariu, T. Thio, L.J. Wang, T.W. Ebbesen, H.J. Lezec: Opt. Lett. **26**, 450 (2001)
37. R. Elghanian, J.J. Storhoff, R.C. Mucic, R.L. Letsinger, C.A. Mirkin: Science **277**, 1078 (1997)
38. P. Bao, A.G. Frutos, C. Greef, J. Lahiri, U. Muller, T.C. Peterson, L. Warden, X. Xie: Anal. Chem. **74**, 1792 (2002)
39. H. Takei: In: *SPIE Microfluidics Devices and Systems* (Santa Clara, California, 1998) 278
40. M. Himmelhaus, H. Takei: Sens. Actuators B **63**, 24 (2000)
41. C.R. Musil, D. Jeggle, H.W. Lehmann, L. Scandella, J. Gobrecht, M. Döbeli: J. Vac. Sci. Technol. B **13**, 2781 (1995)
42. J. Homola, S.S. Yee, G. Gauglitz: Sens. Acutators B **54**, 3 (1999)
43. N.-P. Huang, J. Voros, S.M. De Paul, M. Textor, N.D. Spencer: Langmuir **18**, 220 (2002)
44. A.J. Thiel, A.G. Frutos, C.E. Jordan, R.M. Corn, L.M. Smith: J. Vac. Sci. Technol. B **13**, 2781 (1995)
45. A. Ulman: Chem. Rev. **96**, 1533 (1996)
46. S. Maltis, I. Rubinstein: Langmuir **14**, 1116 (1998)
47. L.S. Jung, C.T. Campbell, T.M. Chinowsky, M.N. Mar, S.S. Yee: Langmuir **14**, 5636 (1998)
48. H. Takei, M. Himmelhaus, T. Okamoto: Opt. Lett. **27**, 342 (2002)
49. Biomolecular Sensors ed. by E. Gizeli, C.R. Lowe (Taylor & Francis, London 2002)

Near-Field Optical-Head Technology for High-Density, Near-Field Optical Recording

T. Matsumoto

1 Introduction

Increased recording density is required in the storage devices used in personal computers, network servers, home servers, and other multimedia devices as users increasingly want to save large amount of images, movies, music files, and so on. To meet this demand, the recording density of storage devices has been steadily increased (Fig. 1). For example, the recording density of optical recording devices reached 19.5 Gb/in^2 in the case of the Blu-ray disk drive introduced to the market in 2003, while that of hard disk drives (HDDs) on the market reached 70 Gb/in^2 in 2002.

However, the rise in recording densities seems to be encountering barriers in both optical and magnetic recording. In the case of optical recording, the recording density depends on the spot size of the focused laser. However, the minimum spot size is limited by the diffraction of light, and the recording

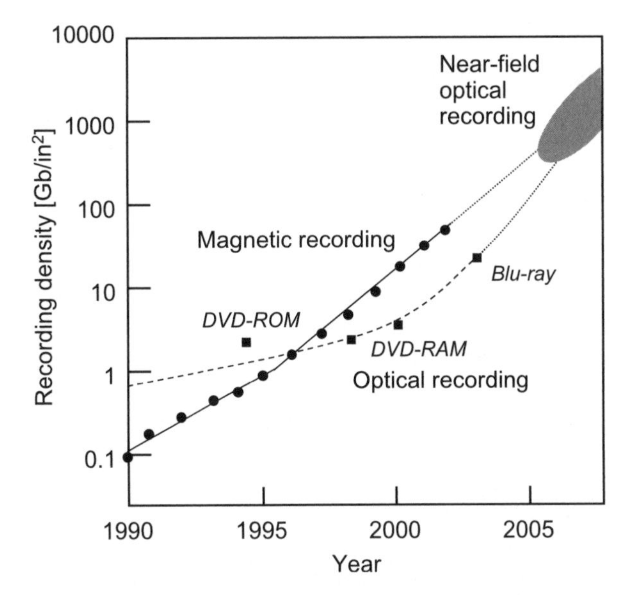

Fig. 1. Growth of the recording density of recording devices

density cannot be increased beyond this limit. In the case of magnetic recording, the recorded information is expected to decay spontaneously because of thermal fluctuation in the magnetic domains when the recording density approaches 1 Tb/in^2. A solution is to increase the coercivity of the magnetic recording medium. However, conventional writing heads cannot generate a magnetic field large enough to write data on such recording medium because of a fundamental limit on the saturation moment of the writing head material.

As a method to overcome these limits, near-field optical recording is now drawing much attention. In this recording, the data bits are written and read by using an optical near field generated near a nanometer-scale object. Because the size of the optical spot is not limited by the diffraction of light, and can be reduced to less than a few tens of nanometers [1], this method is expected to enable a recording density of over 1 Tb/in^2.

In a near-field optical recording device, the key component is a near-field optical head that generates the optical near field. In this chapter, after briefly reviewing near-field optical recording, we describe technical issues regarding the near-field optical head and introduce some solutions to resolve them. Among these issues, improved efficiency in generating the optical near field is the most important. We describe a highly efficient near-field optical head that uses a metallic plate; the intensity of the optical near field is enhanced by a plasmon generated in the metallic plate.

2 Review of Near-Field Optical Recording

2.1 The Limit of Conventional Optical Recording

The recording density of conventional optical recording drives has been increased by reducing the spot size of the focused laser. However, the minimum spot size, D, is governed by the law of diffraction and is approximately

$$D = \lambda/NA\,, \tag{1}$$

where λ is the light wavelength and NA is the numerical aperture of the focusing lens. As an example, in the most advanced optical recording device, the Blu-ray disk drive, the laser wavelength is 405 nm, and the NA is 0.85, thus the spot size is approximately 480 nm. To reduce the spot size further, we must decrease the laser wavelength or increase the NA of the focusing lens. However, a wavelength of less than 405 nm cannot be obtained from currently available semiconductor laser diodes, and it is difficult to make the NA larger than 0.85 because a larger NA reduces the tolerance for recording disk fluctuation. Therefore, it will be difficult to further reduce the spot size.

One way to reduce the spot size is to place a hemispherical or a Weierstrass-sphere shaped lens, called a solid immersion lens (SIL), near the recording medium (Fig. 2) [2, 3]. When an SIL is placed near the recording medium,

	Conventional lens	Solid immersion lens		Near-field optical head (Aperture)
		Hemisphere	Weierstrauss sphere	
Head	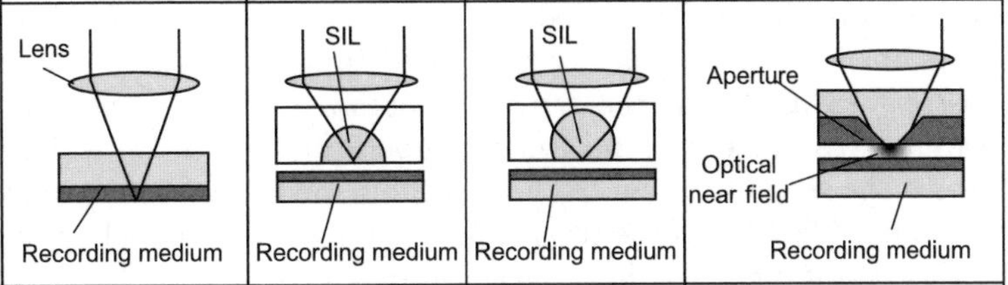 Lens Recording medium	SIL Recording medium	SIL Recording medium	Aperture Optical near field Recording medium
Spot size	λ/NA	$\lambda/(n \cdot NA)$	$\lambda/(n^2 \cdot NA)$	Aperture diameter (<100 nm)
Head media separation	> 1 μm	< 100 nm	< 100 nm	< 10–20 nm

Fig. 2. Light focusing methods in optical recording devices

the wavelength inside the sphere is reduced by a factor of n, where n is the refractive index of the lens; thus, the spot size is reduced to

$$D = \lambda/nNA . \tag{2}$$

In addition, in the case of the Weierstrass-sphere lens, the incident rays are refracted at the sphere's surface. This corresponds to an increased NA, and the spot size is further reduced to

$$D = \lambda/n^2NA . \tag{3}$$

This small optical spot enables increased recording density. For example, a recording density of about 70 $\mathrm{Gb/in}^2$ has been realized using an SIL made of $Bi_4Ge_3O_{12}$ with a refractive index of 2.23 [4]. This type of recording method is sometimes classified as near-field optical recording because it uses an evanescent wave generated by a ray whose incident angle is larger than the critical angle. However, the spot size is still limited by the diffraction limit of light, and no dramatic increase of recording density is expected.

2.2 Near-Field Optical Recording Method

In near-field optical recording, the data bits are written and read by using the optical near field. When light is introduced into a nanometer-scale object, such as a subwavelength aperture or scatterer, a localized electromagnetic field – the optical near field – is generated near the object. The distribution of the optical near field is determined by the object's shape and size, and does not depend on the light wavelength. For example, when the optical near field is generated by the subwavelength aperture, the width of the near-field distribution is as small as the aperture diameter [1]. Therefore, a high recording density beyond the diffraction limit can be realized by writing and reading data bits using the optical near field.

Betzig et al. [5] have demonstrated near-field optical recording using a magneto-optical recording medium. They generated the optical near field using a fiber probe, which is widely used in scanning near-field optical microscopy (SNOM). In the writing process, a Pt/Co multilayer recording medium was heated to the Curie temperature by the optical near field to reverse the magnetization of the recording medium. In the reading process, transmitted light from the recording medium was collected by the fiber probe and the rotation of the light polarization caused by the Faraday effect was detected. They have written and read recording marks with a diameter of 60 nm, which corresponds to a recording density of 170 Gb/in^2.

Near-field optical recording using a phase-change recording medium or a photochromic recording medium has been reported by, respectively, Hosaka et al. [6] and Jiang et al. [7]. In the phase-change recording, the recording medium is heated by light to change its phase from crystalline to amorphous, or vice versa. Hosaka et al. used GeSbTe recording medium, and wrote and read recording marks with a diameter of 60 nm using a fiber probe. In the photochromic recording, the recording marks are written through a photochemical reaction. Jiang et al. wrote and read recording marks with a diameter of 130 nm on a Langmuir–Blodgett film of photochromic material.

2.3 Hybrid Recording Method

A hybrid recording method, known as thermally assisted magnetic recording or heat-assisted magnetic recording, has also been proposed [8, 9]. In this method, data bits are written on a magnetic recording medium by applying a magnetic field while the medium is heated by the optical near field, and read out with a reading head used for hard disk drives, such as a giant magnetoresistive (GMR) head or a tunneling magneto-resistive (TMR) head. Because GMR and TMR heads are highly sensitive to magnetic fields, a high signal-to-noise ratio in the reading signal can be achieved. Note that the problem of thermal fluctuation in the magnetic domains can be solved when data bits are written using the optical near field. During the writing process, the recording medium is heated by light, hence its coercivity is decreased. Therefore, a recording medium with high coercivity can be used to make the magnetic domains stable.

3 Technical Issues Regarding the Near-Field Optical Head

In the primary experiments done on near-field optical recording, a fiber probe was used as the near-field optical head. However, this probe was designed mainly for near-field optical microscopy and is not suitable for practical recording devices. To realize a practical near-field optical head, the following requirements must be met.

1. Precise control of the spacing between the head and the recording medium
2. Integration of peripheral components into the near-field optical head
3. Highly efficient generation of the optical near field

3.1 Precise Control of the Spacing Between the Head and the Recording Medium

The optical near field is localized near the aperture or the scatterer and its intensity decreases exponentially as the distance from the surface of the head increases. Because the decay length is about 10–20 nm, the spacing between the head and the recording medium must be maintained within a range of 10–20 nm.

With the fiber probe, the spacing is controlled by using the shear force acting between the fiber probe and the recording medium [10, 11]. In this method, the fiber probe is dithered parallel to the surface of the recording medium. Because the dithering amplitude is changed by the shear force, a feedback loop can be constructed by detecting the change in the dithering amplitude. However, in a practical recording device, the recording disk rotates at a very high speed. For example, when the recording density is 1 Tb/in^2, and the data transmission rate is 1 Gbps, the linear velocity of the recording medium is about 30 m/s. If the fiber probe is placed near recording medium rotating at such speed, the fiber probe soon crashes because the response based on shear-force feedback is too slow.

To solve this problem, planar near-field optical heads have been proposed. In this type of head, the structure for generating the optical near field is mounted on the slider used for the hard disk drive. In the hard disk drive, the spacing between the head and the recording medium is maintained at a constant value by air pressure acting between the slider and the recording medium. (The air pressure is determined by the design of pads formed on the slider surface.) The head–medium spacing of hard disk drives is currently near 10 nm, and is expected to decrease to less than 10 nm within a few years. By applying this technology to the near-field optical head, we can precisely control and maintain the spacing between the near-field optical head and the recording medium.

Figure 3 shows an example of a planar near-field optical head.

Lee et al. [12] have proposed a silicon planar near-field optical head that has an array of square apertures (Fig. 3a). The 60×60 nm apertures were fabricated by an isotropic etching process with high reproducibility, and a line pattern with linewidths of 250 nm was read out with this near-field optical head.

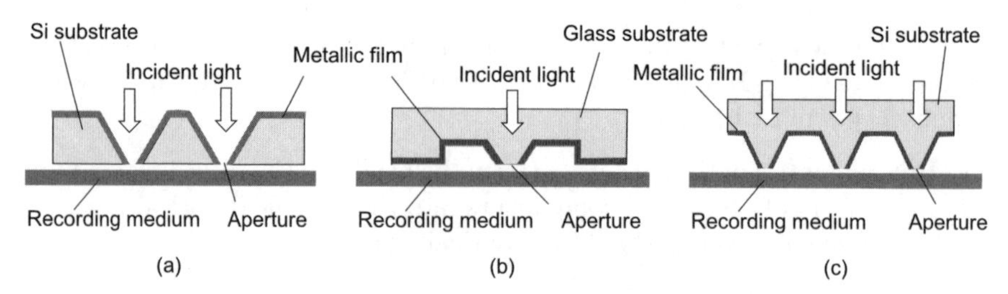

Fig. 3. Examples of planar near-field optical heads: (**a**) Head with an array of apertures formed in a Si substrate, (**b**) Head with an aperture formed on a glass protrusion, (**c**) Head with an array of apertures formed on Si protrusions

Isshiki et al. [13] have proposed a planar near-field optical head with an aperture formed on a pyramidal protrusion (Fig. 3b). They fabricated the pyramidal protrusion by cutting the glass substrate with a focused ion beam, and fabricated the aperture by coating the protrusion with metal and slicing the apex with a focused ion beam. They read out a 250-nm line pattern on a disk rotating at a linear velocity of 0.38 m/s. The data transmission rate and carrier-to-noise ratio were, respectively, 1.5 Mbps and 37 dB (bandwidth: 10 kHz).

Yatsui et al. [14] have also proposed a planar near-field optical head with an array of apertures formed on silicon pyramidal protrusions (Fig. 3c). They fabricated the array of silicon pyramidal protrusions through an isotropic etching process, and formed the apertures at the apices of the protrusions. They used a phase-change recording medium, and succeeded in writing and reading data bits using the apertures. The recording medium was rotated at a linear velocity of 0.43 m/s, and the minimum mark length and the data transmission rate were, respectively, 110 nm and 2.0 MHz. Note that parallel writing and reading using the aperture array is expected to enable a high data transmission rate.

3.2 Integration of Peripheral Components into the Near-Field Optical Head

To realize practical recording devices, several optical components – such as a light source, lens, and detector – must be integrated into a solid near-field optical head. If such components are separated from each other, servo control is needed to adjust the positions of these components. However, such control makes the device very complicated mechanically and electrically. Furthermore, for hybrid recording, integration of a magnetic transducer and magnetic sensor is also required.

Partovi et al. [15] have proposed a near-field optical head where an aperture is formed on a metal-coated facet of the laser diode. They demonstrated writing of recording marks with a size of 250 nm on a phase-change recording medium. In addition, they succeeded in reading out the recording marks by detecting the change in the impedance of the laser caused by the reflected light.

A near-field optical head with subwavelength apertures formed on an array of vertical cavity surface emitting lasers (VCSELs) has been proposed by Goto [16]. A large-scale two-dimensional laser array can be easily fabricated as the VCSEL array. Therefore, this method is advantageous for parallel writing and reading of data bits. In addition, low power consumption is expected because the threshold current is quite low.

Instead of integrating the laser diode into the solid near-field head, light may be delivered to the head by using an optical fiber fixed on the slider. For example, Kato et al. [17] have developed a near-field optical head with an optical fiber fixed horizontally to the slider surface; the light from the optical fiber is guided to an aperture with a micromirror and microlens formed on the slider.

3.3 Obtaining High Efficiency in Generating the Optical Near Field

When a subwavelength aperture is used to generate the optical near field, the efficiency (defined as the ratio between the power of the optical near field and that of the incident light) is quite low. For example, the efficiency is 10^{-4}–10^{-5} when the aperture diameter is 60 nm [5], and it decreases further as the aperture diameter becomes smaller. This is because a cutoff diameter exists for mode propagation in the waveguide. When the diameter of the tapered waveguide becomes smaller than the cutoff diameter, the propagating light suffers a huge loss and its intensity decays exponentially.

This low efficiency is a serious hindrance to realizing practical near-field optical recording devices. For example, in the case of hybrid recording, the temperature rise at the heated point in the recording medium is 40°C when a recording medium moving at a linear velocity of 15 m/s is irradiated by light with a spot size of 50 nm, peak power of 1 mW, and a pulsewidth of 2 ns [18]. Given these values and the output power of available laser diodes, the required efficiency for writing is 1–10% if we assume that the temperature rise required for writing is 200°C.

To raise efficiency, several methods have been proposed, such as

1. Optimization of the waveguide shape
2. Use of a surface plasmon
3. Use of a metallic waveguide
4. Use of a planar antenna

Low efficiency has also been a problem for SNOM when observing a weak optical signal, such as photoluminescence, or a nonlinear signal. Although most of these methods have been developed for SNOM applications, they may also be applied to near-field optical recording. Here, we explain these methods briefly.

Optimization of the Waveguide Shape

A double- and triple-tapered fiber probe (Fig. 4) have been proposed to increase the efficiency by, respectively, Saiki et al. [19] and Yatsui et al. [20]. In this type of fiber probe, efficiency is increased by optimizing the waveguide shape. As mentioned, when light is introduced into the tapered waveguide, the light power decreases in the loss region where the waveguide diameter is smaller than the cutoff diameter. In double- and triple-tapered fiber probes, high efficiency is realized by shortening the length of the loss region. In the case of the triple-tapered probe, efficiency is also increased because the edge

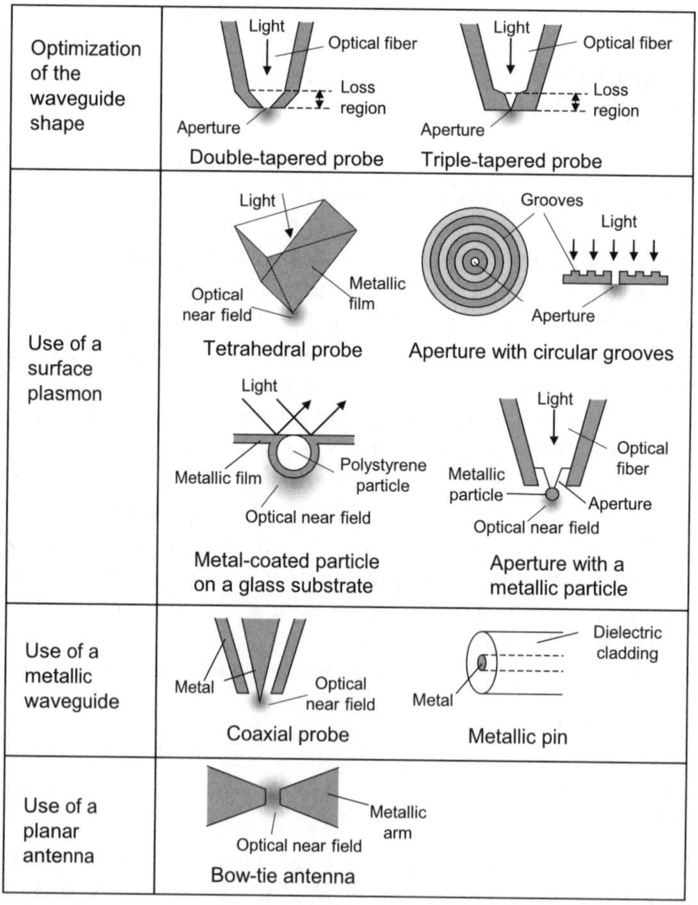

Fig. 4. Examples of methods to increase efficiency

between the second and third tapers excites the HE-plasmon mode, which has a smaller cutoff diameter. The efficiency of the double-tapered probe is 10 times higher than that of a conventional probe, and the efficiency of the triple-tapered probe is 1000 times higher than that of a conventional probe.

Use of a Surface Plasmon

Fischer et al. [21, 22] have proposed a tetrahedral probe that uses the surface plasmon generated on a surface of a metallic film. The surface plasmon is a collective oscillation of electrons generated on a surface of a thin metallic film by irradiating the film with light [23]. When the surface plasmon is excited, the electric field near the metallic field is enhanced. In the tetrahedral probe, the surface plasmon is excited on the metallic film on the tetrahedral glass chip, and propagates toward the apex. Thus, the electric field at the apex is enhanced.

Thio et al. [24] have proposed a structure that consists of a subwavelength aperture created in a metallic film and a set of concentric circular grooves surrounding the aperture. When light is introduced into this structure, the surface plasmon is excited on the metallic film by the circular grooves, and the optical near-field intensity at the aperture is enhanced because the surface plasmon assists the light transmitting through the aperture. In their experiment, the transmission of the aperture with the grooves was 50 times that of an aperture without grooves.

Fischer and Pohl [25] have also proposed a probe that uses a surface plasmon generated at a subwavelength metallic sphere.

When light is introduced into the subwavelength metallic sphere, the dipole moment p inside the sphere is given by

$$p = 4\pi\varepsilon_0 a^3 \frac{\varepsilon(\omega) - 1}{\varepsilon(\omega) + 2} E_0 , \tag{4}$$

where $\varepsilon(\omega)$ is the dielectric constant of the sphere, ω is the angular frequency of the light, a is the radius of the sphere, ε_0 is the dielectric constant of vacuum, and E_0 is the magnitude of the applied electric field. The field E_s at the surface of the sphere is the superposition of the field generated by the dipole and the applied field E_0, and is given by

$$E_\mathrm{s} = 3 \frac{\varepsilon(\omega)}{\varepsilon(\omega) + 2} E_0 . \tag{5}$$

As shown in this equation, E_s is enhanced when the light frequency satisfies the equation

$$\mathrm{Re}[\varepsilon(\omega)] = -2 . \tag{6}$$

As an example, Fig. 5 shows the relation between the wavelength and an enhancement factor T given by

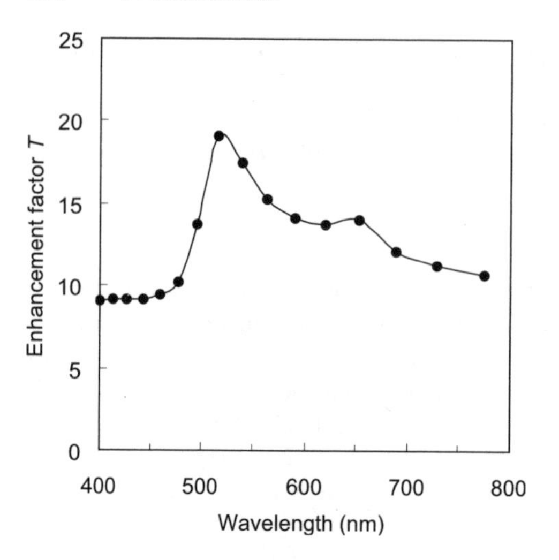

Fig. 5. Relation between the enhancement factor and the wavelength for a gold sphere calculated by a dipole model

$$T = |E_\mathrm{s}/E_0|^2 \tag{7}$$

for a gold sphere. In this calculation, the experimental value given in [34] was used as the dielectric constant of the gold. As shown in this figure, the field at the surface became greater when the wavelength was near 500 nm.

The peak corresponds to a resonance condition of the collective oscillation of the electrons in the metallic sphere. When the sphere is illuminated by light, charges in the sphere move to the surface (Fig. 6). These charges give rise to an internal field E_i, hence a restoring force acts on the charges. This force causes a charge oscillation that has a resonance frequency satisfying (6). If the light frequency corresponds to the resonance frequency of the charge oscillation, light energy is strongly absorbed by the sphere and the optical near field at the surface is enhanced. The quantum of the collective oscillation of the charges in the small metallic particle is called a localized surface plasmon or a localized plasmon.

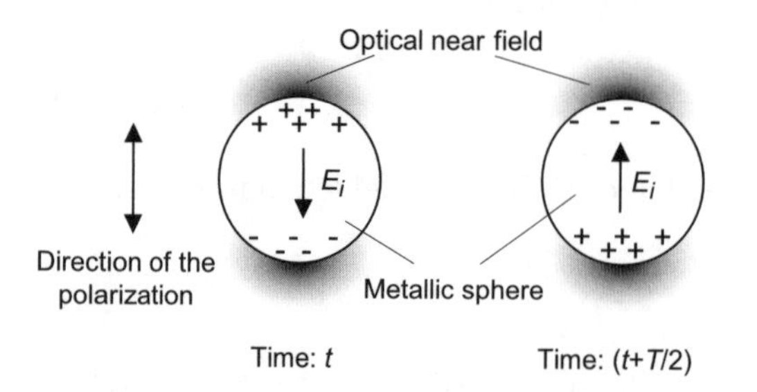

Fig. 6. Schematic of a localized plasmon excited in a metallic particle

In the experiment by Fischer and Pohl, a small polystyrene sphere coated with gold film was formed on a prism surface, and it was illuminated by light under a total internal reflection condition to excite the localized plasmon. The field-enhancement effect by the localized plasmon was observed by placing a substrate near the particle and detecting the scattered light caused by the interaction with the substrate.

A fiber probe having a metallic particle at its apex has also been proposed by Matsumoto et al. [26]. An aperture was formed at the apex of a tapered optical fiber, and the particle was created at the center of the aperture. Because the light was introduced and collected through the aperture, the amount of background light in the detected light was reduced.

Use of a Metallic Waveguide

A coaxial probe and a metallic pin probe have been proposed by, respectively, Fischer [27] and Takahara et al. [28]. These probes have metallic cores, and high efficiency is expected because there exists a mode propagating at the surface of the metallic core, and this mode does not have a cutoff diameter.

Use of a Planar Antenna

Grober et al. [29] have proposed the use of a bow-tie antenna, which has been used for microwave applications, as an optical near-field generator. When an electromagnetic wave is incident upon such an antenna, current is induced in the two arms, and charges accumulate at the apices. These charges work as a dipole and a strong electromagnetic field is generated at the gap. To check the feasibility of this approach, they used microwaves with a wavelength of 13.6 cm (2.2 GHz), and used a dipole antenna to observe the electromagnetic field localized at the gap.

4 Novel Design of a Near-Field Optical Head Using a Plasmon

To realize a highly efficient near-field optical head, we have developed a near-field optical head using a plasmon generated by a wedge-shaped metallic plate [30, 31]. In this section, we describe its principle, simulation results, and its fabrication results.

4.1 Principle

This head consists of a transparent substrate and a wedge-shaped metallic plate – a metallic plate in the shape of a sector or a triangle, formed at the bottom of the substrate (Fig. 7). When the metallic plate is illuminated by

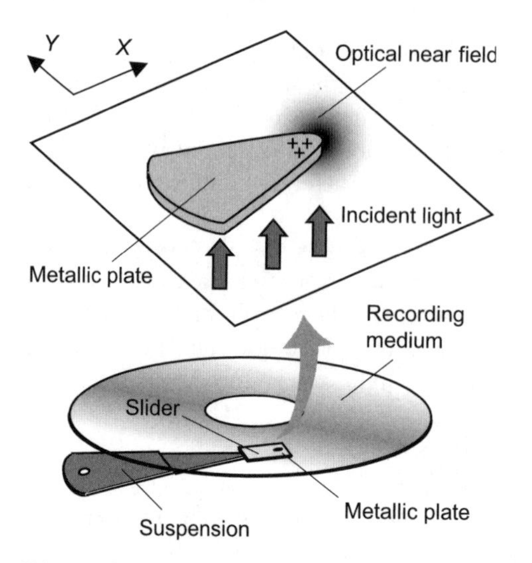

Fig. 7. Schematic of the near-field optical head with a wedge-shaped metallic plate

light polarized in the x direction, charges oscillate in the x direction, and concentrate at the apex. These concentrated charges generate a localized electromagnetic field: i.e., an optical near field close to the apex of the metallic plate.

The optical near field is enhanced when the plasmon is excited in the metallic plate. That is, the displacement of charges creates a depolarization field inside the metallic plate. Because of this field, the charges act like an oscillator system as is the case with a metallic sphere. When the light frequency corresponds to the resonance frequency of this oscillation, light energy is strongly absorbed and the intensity at the apex is enhanced.

This type of near-field optical head has the following advantages:

1. High efficiency
   The field-enhancement effect of the localized plasmon enables generation of a strong optical near field. Although the localized plasmon is also generated in the metallic sphere, a stronger optical near field can be generated by using the wedged-shaped metallic plate as described in Sect. 4.4.
2. Simple structure
   Because it is a planar structure, the head can be easily fabricated by techniques used in the fabrication of semiconductor devices, such as electron beam lithography. This is beneficial for integration with other components such as the laser diode and the lens. Furthermore, if the metallic plate is embedded in the slider surface as described in Sect. 4.7, high durability can be realized because there is no protruding structure.

4.2 Simulation Method

To check the feasibility of the proposed head, the distribution of the optical near field was calculated through a finite difference time-domain (FDTD) calculation [32, 33].

In this calculation, the incident light was assumed to be a plane wave. Experimental values given in [34] were used as the dielectric constants of the metal. The cell size was made large near the boundary and small near the object to be calculated to minimize the computer memory required. When the object was the wedge-shaped metallic plate, the cell size was designed to have the smallest value near the apex, and the minimum cell size was 1/10 of the radius of the apex. As the boundary condition, Mur's absorption boundary condition was used. A parallel computer with 24 processors was used for the calculation.

4.3 Use of an Aperture and a Circular Metallic Plate

To begin with, for reference we will show the distribution generated near a subwavelength aperture and a circular metallic plate.

Aperture

Figure 9 shows the intensity distribution of an optical near field generated near an aperture created in a metallic film as shown in Fig. 8. For this calculation, we assumed that the metallic film was a gold film with a thickness t of 100 nm, the diameter a of the aperture was 30 nm, and the wavelength of the incident light was 650 nm. The polarization of the incident light was in the direction of the arrows in Fig. 9. The distance d between the observation plane and the aperture was 5 nm. The unit of intensity was the ratio between the intensity of the optical near field and that of the incident light. As shown in this figure, the intensity at the aperture became lower than the incident light intensity. The intensity at the peak was about 0.04 times that of the incident light.

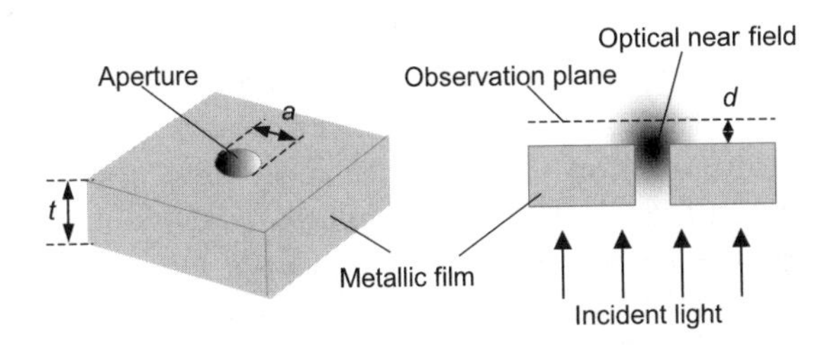

Fig. 8. Simulation model for an aperture

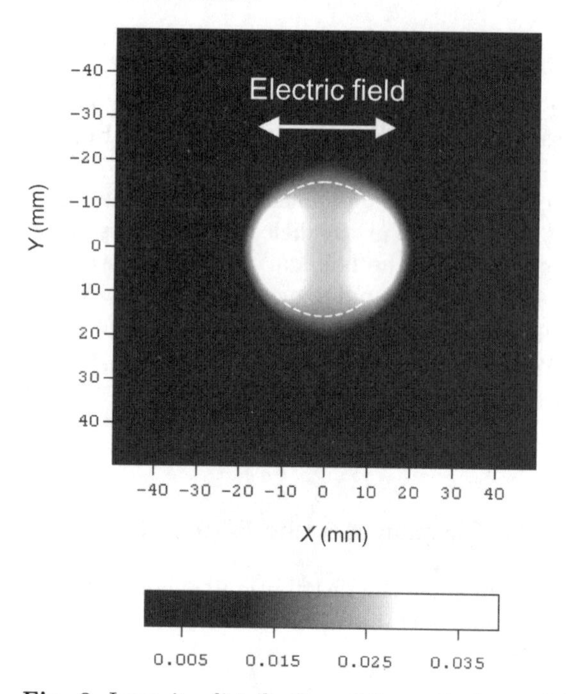

Fig. 9. Intensity distribution of the optical near field generated near an aperture created in a metallic film. It was assumed that the metal was gold, the aperture diameter was 30 nm, and the thickness of the metallic film was 100 nm. The distribution was calculated on a plane 5 nm from the metallic plate

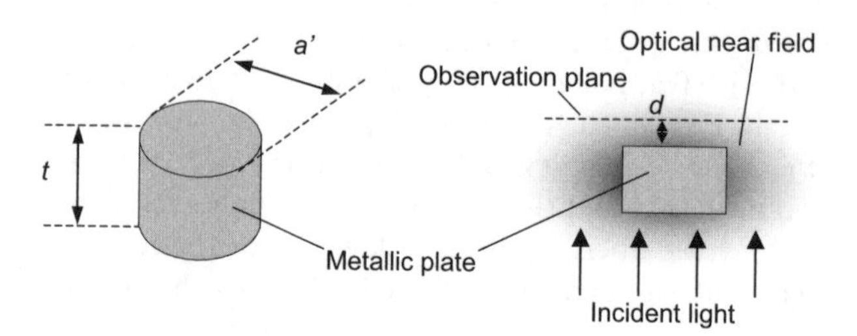

Fig. 10. Simulation model for a circular metallic plate

Circular Metallic Plate

Figure 11 shows the intensity distribution of the optical near field near a circular metallic plate as shown in Fig. 10. For this calculation, we assumed that both the diameter a' and the thickness t were 30 nm, the metal was gold, and the wavelength of the incident light was 650 nm. The polarization of the incident light was in the direction of the arrows shown in Fig. 11. The distance d between the observation plane and the aperture was 5 nm. As shown in this figure, the near-field intensity became high at the edges A

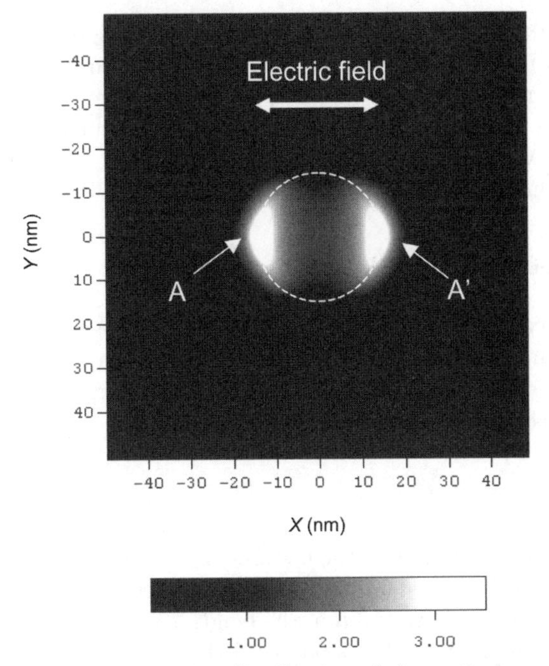

Fig. 11. Intensity distribution of the optical near field generated near a circular metallic plate. It was assumed that the metal was gold, and the circle diameter and the metallic film thickness were both 30 nm. The distribution was calculated on a plane 5 nm from the metallic plate

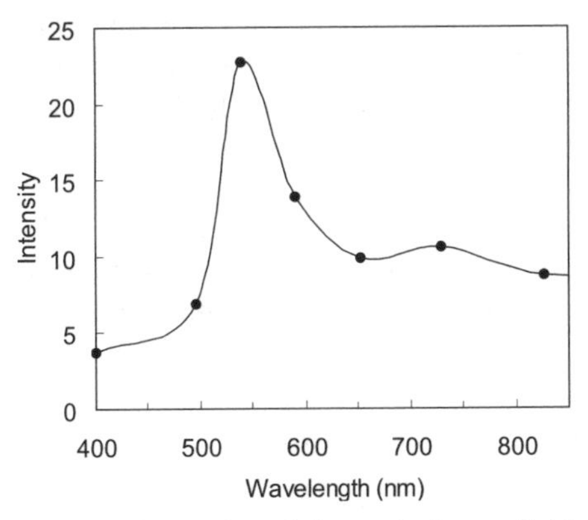

Fig. 12. Intensity of the optical near-field generated near the circular metallic plate as a function of the incident light wavelength. The intensity was calculated on a plane 2 nm from the metallic plate

and A′ of the metallic plate. This is because the charges in the metallic plate accumulated at these edges. The intensity at the peak was 3.5 times that of the incident light. Figure 12 shows the dependence of the optical near-field intensity (intensity at the peak) on the wavelength of the incident light. In this calculation, the near-field intensity was calculated on a plane 2 nm from the metallic plate. As shown in this figure, the curve was almost the same as that of the metallic sphere calculated by (7), and the optical near-field intensity became high near a wavelength of 550 nm, which corresponds to the resonance wavelength of the localized plasmon.

4.4 Use of a Wedge-Shaped Metallic Plate Placed in Air

In this subsection, we will present our simulation results for the wedge-shaped metallic plate placed in air, and describe the fundamental properties of the optical near field generated near the metallic plate, such as the intensity distribution, efficiency, and resonance property of the plasmon.

Near-Field Distribution

Figure 14 shows the near-field distribution near the metallic plate in the shape of a sector as shown in Fig. 13a. For this simulation, we assumed that the metal was gold, the radius r of the apex was 10 nm, the length l was 150 nm, the thickness t was 30 nm, the apex angle θ was 60°, and the wavelength of the incident light was 690 nm. The distribution was calculated on a plane 5 nm from the metallic plate. The unit of intensity was the ratio between the intensity (power density) of the optical near field and that of the incident light. As shown in this figure, a strong optical near field was generated at the apex, and the peak intensity was 350 times that of the incident light. This intensity was 8750 times that of the 30-nm aperture created in the metallic film, and 85 times that of the circular metallic plate. The distribution measured at half-maximum was 19 nm wide in the x direction and 25 nm wide in the y direction.

For this type of near-field optical head, the polarization must be in the x direction as shown in Fig. 13. Figure 15 shows a near-field distribution when the polarization of the incident light was in the y direction. As shown

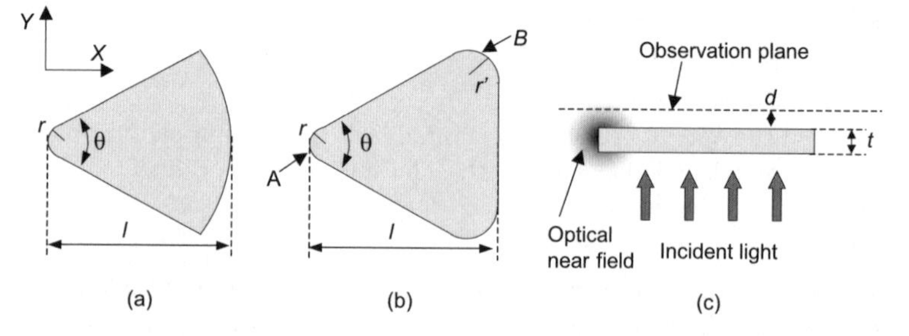

Fig. 13. Simulation model for the wedge-shaped metallic plate: (**a**) metallic plate in the shape of a sector, (**b**) metallic plate in the shape of a triangle, (**c**) side view

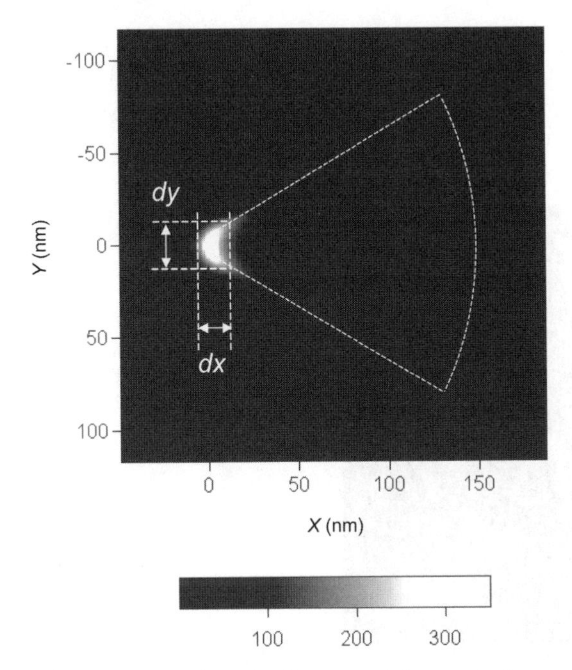

Fig. 14. Intensity distribution of the optical near field generated near a metallic plate in the shape of a sector. It was assumed that the radius r of the apex was 10 nm, the length l was 150 nm, the thickness t was 30 nm, the apex angle θ was 60°, and the incident light wavelength was 690 nm. The unit of intensity was the ratio between the intensity of the optical near field and that of the incident light

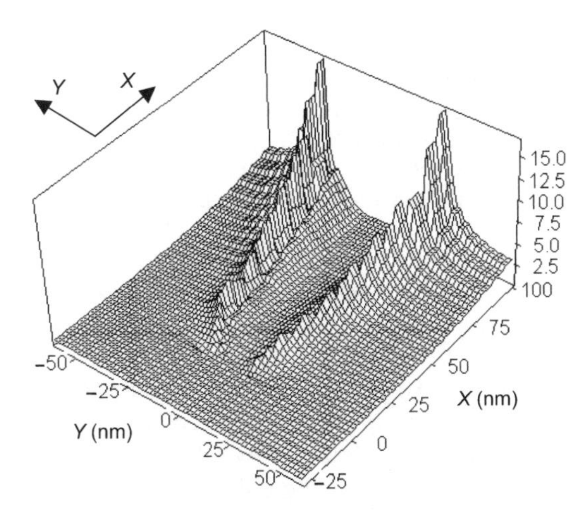

Fig. 15. Intensity distribution of the optical near field when the polarization was in the y direction

in this figure, the optical near field became strong along the side edges of the metallic plate, but weak at the apex. This is because the charges oscillate in the direction parallel to the y axis; hence the charges accumulate along the side edges instead of accumulating at the apex.

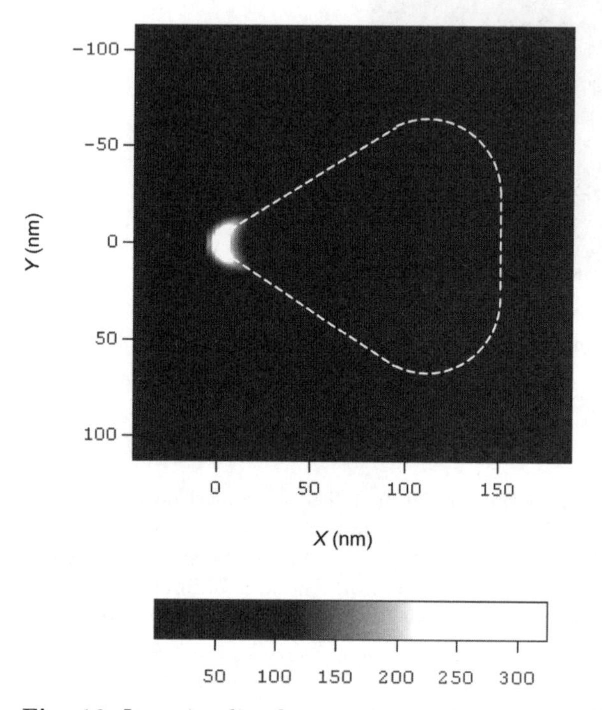

Fig. 16. Intensity distribution of the optical near field generated near a metallic plate in the shape of a triangle. It was assumed that the metal was gold, the radius r of apex A was 10 nm, the radius r' of apices B was 30 nm, the apex angle θ was 60°, the thickness t was 30 nm and the incident light wavelength was 690 nm. The distribution was calculated on a plane 5 nm from the metallic plate

Figure 16 shows the near-field distribution when the shape of the metallic plate was a triangle. For this calculation, we assumed that the metal was gold, the radius r of apex A was 10 nm, the radius r' of apices B was 30 nm, the apex angle θ was 60°, the thickness t was 30 nm and the wavelength of the incident light was 690 nm. The distribution was calculated on a plane 5 nm from the metallic plate. As shown in this figure, a strong optical near-field was also generated at apex A as with the metallic plate in the shape of the sector. The distribution at the apex was the same as that of the metallic plate in the shape of the sector. Note that the radius r' of apices B should be larger than the radius r of the apex A, otherwise charges also concentrate at apices B, and a high optical near-field intensity is also generated there.

Dependence of the Spot Size on the Distance from the Metallic Plate

Because the optical near field is localized near the metallic plate, the size of the optical spot depends on the distance from the metallic plate d. Figure 17 shows the spot size as a function of the distance from the metallic plate for a metallic plate in the shape of the sector, and Fig. 18 shows the distributions for distances of 2, 5, and 10 nm. In Fig. 17, the solid line represents the width measured at half-maximum in the x direction dx, and the broken line represents the width measured at half maximum in the y direction dy. As shown, the spot size decreased as the distance became smaller. For example, the spot size was d$x = 9$ nm, d$y = 16$ nm when the distance was 2 nm, while it was d$x = 28$ nm, d$y = 30$ nm when the distance was 10 nm.

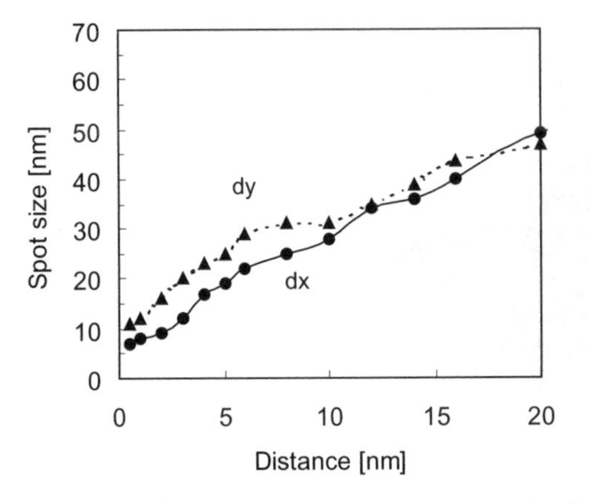

Fig. 17. Dependence of the spot size on the distance from the metallic plate when the radius of the apex was 10 nm

Dependence of the Spot Size on the Radius of the Apex

The size of the optical spot also depends on the radius r of the apex. Figure 19 shows the size of the optical spot as a function of the radius of the apex simulated for the metallic plate in the shape of the sector. The solid line shows the width measured at half-maximum in the x direction dx, and the broken line shows the width measured at half-maximum in the y direction dy. The distribution was calculated on a plane 5 nm from the metallic plate. As shown, the spot size decreased linearly as the radius of the apex became smaller.

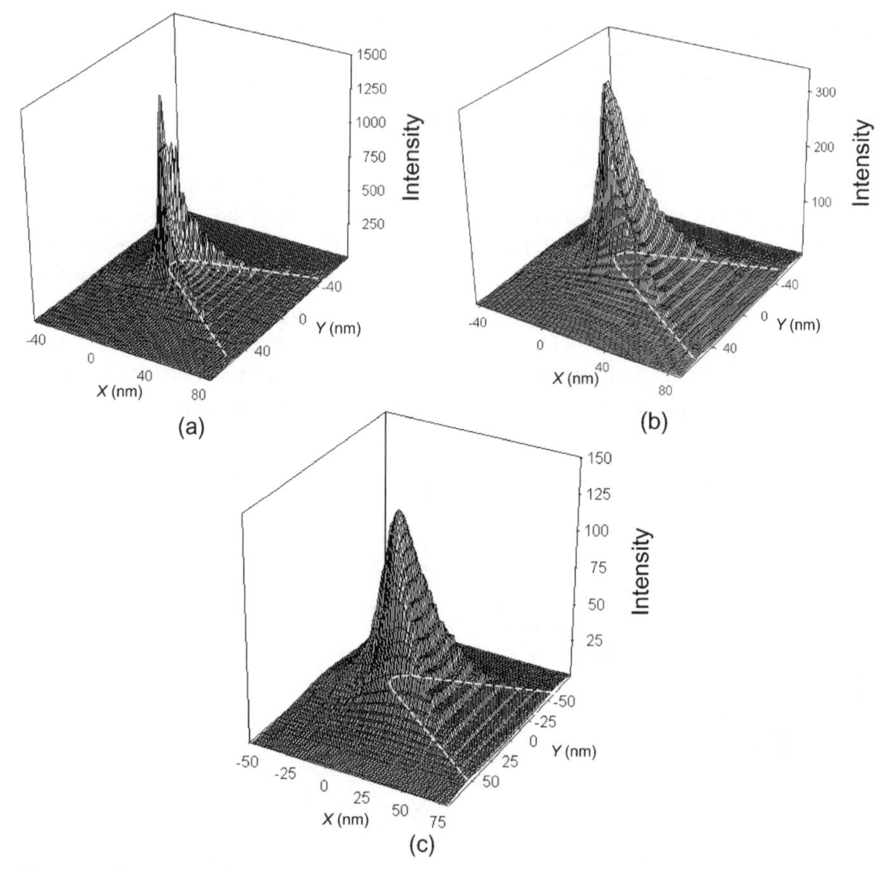

Fig. 18. Intensity distribution of the optical near field generated near a metallic plate in the shape of a sector for a distance of (**a**) 2 nm, (**b**) 5 nm, (**c**) 10 nm

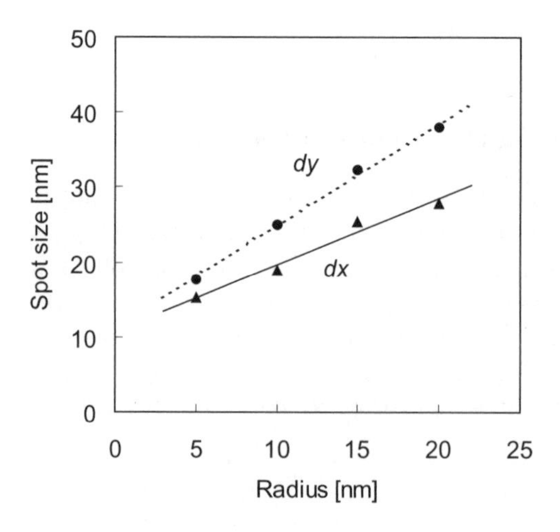

Fig. 19. Dependence of the spot size on the radius of the apex. The distance between the observation plane and the metallic plate was 5 nm

Efficiency

Here, we estimate the efficiency η by defining it as

$$\eta = \frac{\int_{\mathrm{S}} p_{\mathrm{near}} \mathrm{d}S}{\int_{\mathrm{S}'} p_{\mathrm{in}} \mathrm{d}S'} , \tag{8}$$

where p_{near} is the power density (intensity) of the optical near field, p_{in} is that of the incident light, and S and S' are the areas where the power density is higher than half-maximum. The efficiency calculated for the near-field distribution in Fig. 14 was about 15% if we assume that the incident beam was a Gaussian beam with a full width at half-maximum of 1 μm (corresponding to an optical spot focused by a lens with an NA of 0.35). When we calculated the near-field distribution, we assumed the incident light was a plane wave. Therefore, in this estimation, the diameter of the incident beam was assumed to be 1 μm so that the incident light intensity would be nearly constant at the metallic plate. However, under such a condition, the energy loss is large because much of the light is not incident on the metallic plate. Higher efficiency should be attainable by reducing the diameter of the incident light.

In the case of the aperture, the near-field intensity became lower, and the efficiency decreased rapidly as the optical spot became smaller. However, for the head using the metallic plate, the near-field intensity increased when the spot size became smaller. As a consequence, the efficiency did not decrease as rapidly as the aperture. The solid line in Fig. 20 shows the intensity as a function of the apex radius simulated for the metallic plate in the shape of the sector. The intensity represents the peak intensity in a distribution calculated on a plane 5 nm from the metallic plate. As shown, the near-field intensity rose as the apex radius became smaller. This is because the oscillating charges concentrated in a smaller area, and the charge density increased when the apex radius became smaller. The broken line in Fig. 20 shows efficiency as a function of the apex radius. The efficiency was about 10% even if the radius was 5 nm (corresponding to an optical spot 15 nm by 18 nm).

Tuning of the Resonance Wavelength of the Plasmon

The wavelength of the incident light must correspond to the resonance wavelength of the plasmon to generate a strong optical near field. However, the available wavelength is limited if we use a semiconductor laser as the light source. Therefore, we need to tune the resonance wavelength of the plasmon to the wavelength of the semiconductor laser. For the wedge-shaped metallic plate, the resonance wavelength of the plasmon can be adjusted by changing the material or the length l of the metallic plate. In this subsection, we will describe how the resonance wavelength depends on these parameters.

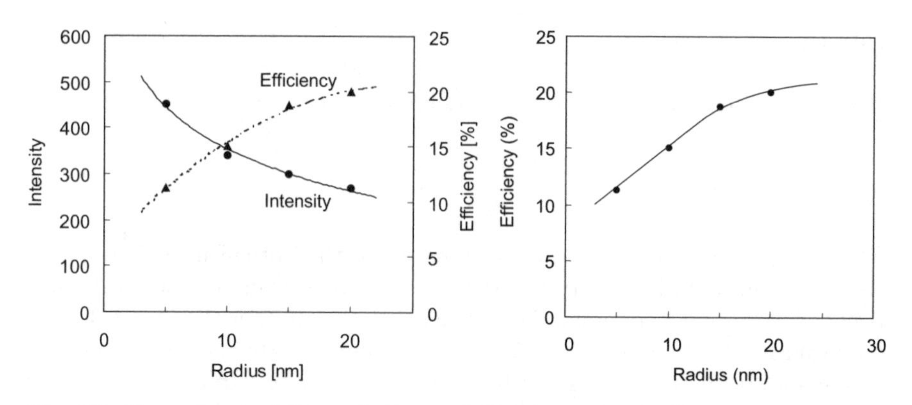

Fig. 20. Dependence of the optical near-field intensity and efficiency on the apex radius. It was assumed that the length l, the apex angle θ, and the incident light wavelength were 150 nm, 60°, and 690 nm, respectively. The intensity represents the peak intensity in a distribution measured on a plane 2 nm from the metallic plate. For the calculation of efficiency, the incident light was assumed to be a Gaussian beam with a full width at half-maximum of 1 μm

Dependence on the Material

Figure 21 shows the relation between the optical near-field intensity and the incident light wavelength for metallic plates made of gold, silver, and aluminum. For this calculation, we assumed the shape of the metallic plate was a sector with an apex radius r of 20 nm, a length l of 100 nm, a thickness

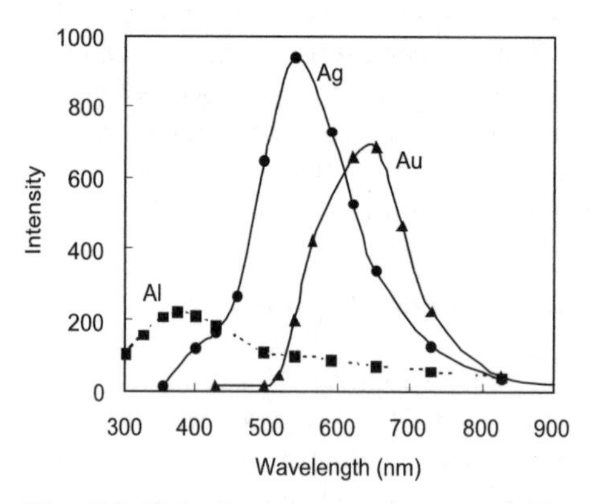

Fig. 21. Relation between the near-field intensity and the wavelength for metallic plates made of silver, gold, and aluminum. The shape of the metallic plate was assumed to be a sector with an apex radius r of 20 nm, a length l of 100 nm, a thickness t of 30 nm, and an apex angle θ of 60°. The closed triangles, closed circles, and closed squares represent the values for gold, silver, and aluminum, respectively. The intensity represents the peak intensity in the distribution calculated on a plane 2 nm from the metallic plate

t of 30 nm, and an apex angle θ of 60°. The intensity represents the peak intensity in a distribution calculated on a plane 2 nm from the metallic plate. The closed triangles, closed circles, and closed squares represent the values for gold, silver, and aluminum, respectively. The resonance wavelength of gold was about 650 nm, which is near the wavelengths of a red semiconductor laser (635 nm, 650 nm, and 670 nm). The resonance wavelength of aluminum was 370 nm, which is near the wavelength of a blue semiconductor laser (405 nm). In the case of silver, the resonance wavelength was 550 nm. Although, there is no semiconductor laser of such wavelength, the resonance wavelength can be shifted toward the red or near-infrared regions by changing the length l as we explain next.

The near-field intensity at the resonance wavelength became larger in the order Al $<$ Au $<$ Ag. This indicates that the near-field intensity becomes large when the imaginary part of the dielectric constant becomes small. This is because the damping of the charge oscillation, which causes energy loss, becomes small when the imaginary part of the dielectric constant is small. To realize high efficiency, we therefore need to use a metal whose dielectric constant has a small imaginary part.

Dependence on the Length of the Metallic Plate

Figure 22 shows the resonance curves when the length l was varied from 100 nm to 600 nm. For this calculation, we assumed that the shape of the metallic plate was a sector with an apex radius r of 20 nm, a thickness t of 30 nm, and an apex angle θ of 60°, and that the metal was silver. The intensity

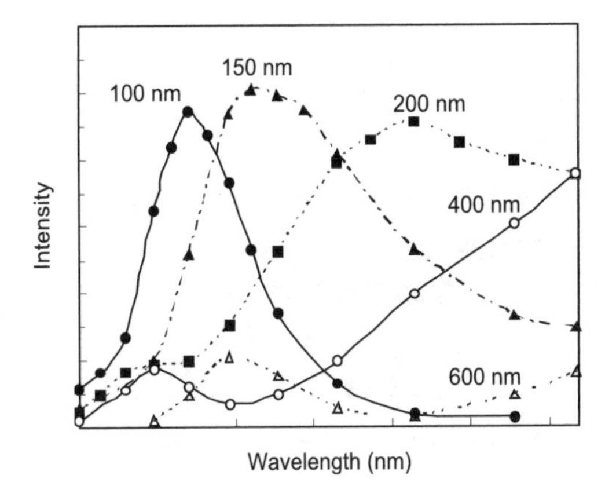

Fig. 22. Relation between the near-field intensity and the wavelength for metallic plates of various lengths. It was assumed that the shape of the metallic plate was a sector with an apex radius r of 20 nm, a thickness t of 30 nm, and an apex angle θ of 60°, and that the metal was silver. The intensity represents the peak intensity in the distribution calculated on a plane 2 nm from the metallic plate

represents the peak intensity in the distribution calculated on a plane 2 nm from the metallic plate. As shown, the resonance wavelength of the plasmon shifted toward a longer wavelength when the metallic plate became longer. We expected the restoring force acting on the charges to become weaker when the plate became longer, and so the resonance frequency would decrease. When the length became greater than 400 nm, a second peak appeared at a short wavelength. This was because another resonance mode was excited at this peak. The broadening of the resonance curve when the length increases was probably due to the radiation damping [35].

4.5 Use of a Wedge-Shaped Metallic Plate Placed near the Recording Medium

In the previous calculation, the metallic plate was placed in air. However, in recording devices, the metallic plate is formed on the slider (a transparent substrate), and placed near the recording medium. In this subsection, we will describe the resonance property of the plasmon and the near-field distribution in such a situation.

Simulation Model

To realize high durability, the metallic plate should be embedded in the slider surface as shown in Fig. 23. Thus, we assumed that the metallic plate was embedded in a glass substrate with a refractive index of 1.45. The recording medium was assumed to be a thin film of TbFeCo or GeSbTe with a thickness of 20 nm, and the separation between the recording medium and the metallic plate was assumed to be 10 nm. For the metallic plate, we assumed that the metal was gold, the apex radius r was 10 nm, the length l was 150 nm, the thickness t was 30 nm, and the apex angle θ was 60°.

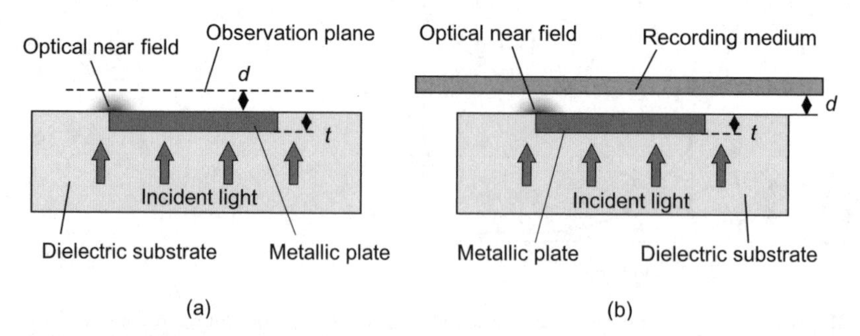

Fig. 23. Simulation model when the recording medium and slider were placed near the metallic plate: (**a**) when there was no recording medium, (**b**) when there was a recording medium

Resonance Property of the Plasmon

First, we show the influence of the glass substrate on the resonance property of the plasmon. Figure 24 shows the resonance curve when there was no recording medium. The solid line represents the resonance curve for the metallic plate embedded in the glass substrate and the broken line represents the resonance curve when the metallic plate was placed in air. The intensity represents the peak intensity in the distribution calculated on a plane 2 nm from the metallic plate. As shown, the resonance wavelength shifted by about 50 nm toward a longer wavelength when the metallic plate was embedded in the glass substrate. We believe this was because the dipoles in the dielectric substrate interact with the charges in the metallic plate, and weaken the restoring force working on the charges.

Figure 25 shows the resonance curves when the recording medium was placed near the metallic plate embedded in the glass substrate. The solid line represents the resonance curve when the TbFeCo recording medium was used, the broken line represents the resonance curve when the GeSbTe medium was used, and the dash-dot line represents the resonance curve when there was no recording medium. The intensity represents the peak intensity in the distribution calculated on a plane 2 nm from the metallic plate. As shown, the resonance wavelength was shifted toward a longer wavelength by interaction with the recording medium. The shift was about 50 nm for both TbFeCo and GeSbTe.

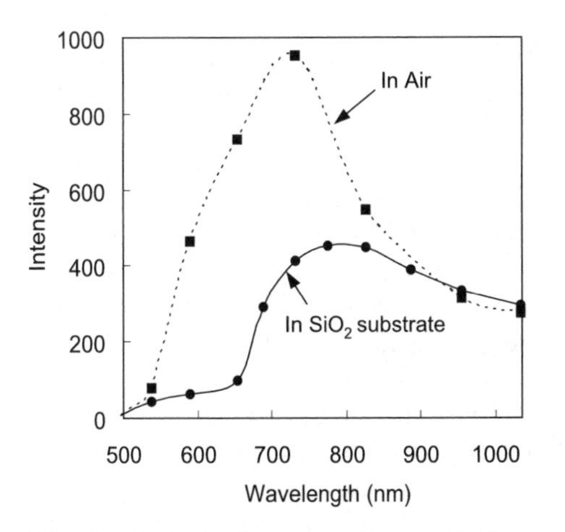

Fig. 24. Relation between the near-field intensity and the wavelength for the metallic plates embedded in a glass substrate. The *solid line* represents the resonance curve for the metallic plate embedded in the glass substrate and the *broken line* represents the resonance curve when the metallic plate was placed in air. The intensity represents the peak intensity in the distribution calculated on a plane 2 nm from the metallic plate

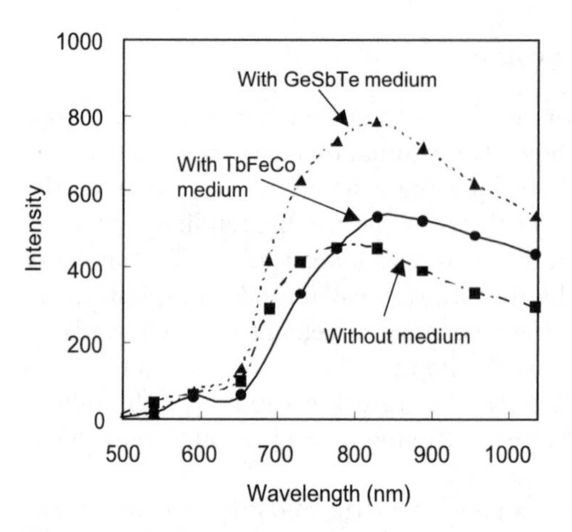

Fig. 25. Relation between the near-field intensity and the wavelength when a recording medium was placed near the metallic plate. The *solid line* represents the resonance curve for a TbFeCo recording medium, the *broken line* represents the resonance curve for a GeSbTe recording medium, and the *dash-dot* line represents the resonance curve when there was no recording medium. The separation between the recording medium and the metallic plate was assumed to be 10 nm. The intensity represents the peak intensity in the distribution calculated on a plane 2 nm from the metallic plate

Near-Field Distribution

Figure 26 shows the intensity distribution of the optical near-field when there was no recording medium, and Fig. 27 shows the intensity distribution when the TbFeCo recording medium was placed near the metallic plate. The wavelength of the incident light was assumed to be at the resonance wavelength of the plasmon – 780 nm when there was no recording medium, and 830 nm when there was a recording medium. Both distributions were calculated on a plane 10 nm from the metallic plate (on the surface of the recording medium when a recording medium was used). As shown in these figures, the optical spot was semicircular when the recording medium was used, but was crescent-shaped when there was no recording medium. With the recording medium, the distribution measured at half-maximum was about 30 nm wide in both the x and y directions. If we compare this spot size with that of a 4.7-GB, DVD-RAM (full width at half-maximum of the optical spot = 580 nm, recording density = 3.7 Gb/in^2), this spot size corresponds to a recording density of approximately 1 Tb/in^2. The calculated efficiency for this distribution was 20% when we assumed that the incident beam was a Gaussian beam with a full width at half-maximum of 1 μm.

When we placed the recording medium near the metallic plate, the intensity of the optical near field increased by a factor of about 2.5. This was because the decay length of the optical near field changes when the recording

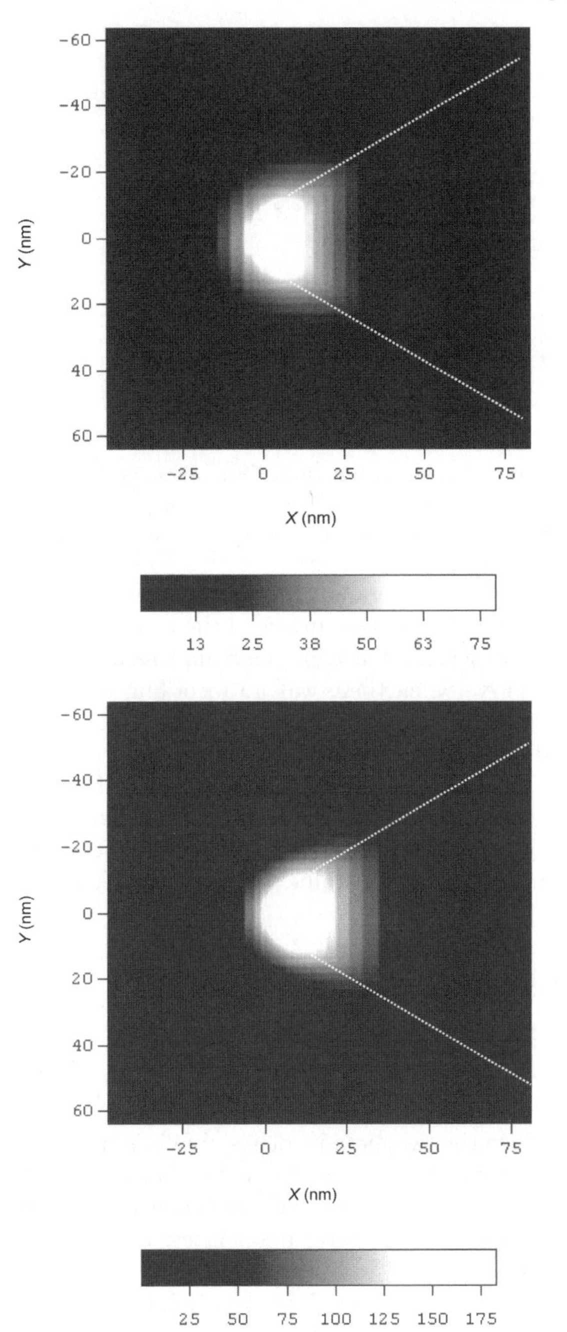

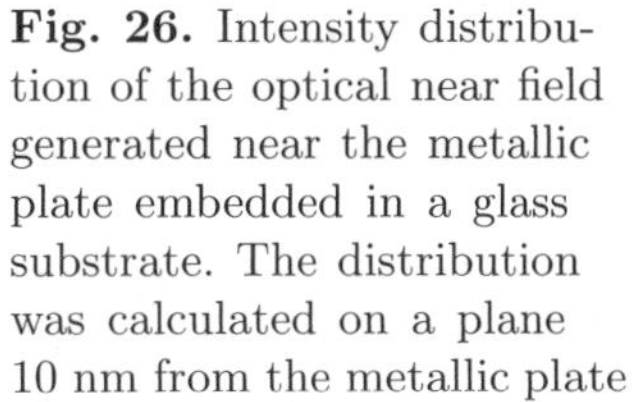

Fig. 26. Intensity distribution of the optical near field generated near the metallic plate embedded in a glass substrate. The distribution was calculated on a plane 10 nm from the metallic plate

Fig. 27. Intensity distribution of the optical near field when a TbFeCo recording medium was placed near the metallic plate. The distribution was calculated on a plane 10 nm from the metallic plate (on the surface of the recording medium)

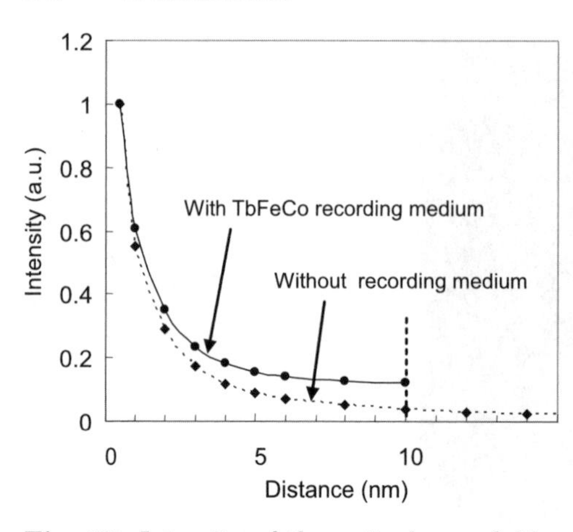

Fig. 28. Intensity of the optical near field as a function of the distance from the metallic plate. The *solid line* represents the curve when a recording medium was used, and the *broken line* represents the curve when there was no recording medium

medium is placed near the metallic plate. Figure 28 shows the dependence of the near-field intensity on the distance from the surface of the metallic plate. The solid line represents the curve when a recording medium was used, and the broken line represents the curve when there was no recording medium. The near-field intensity decreased exponentially as the distance increased, and its decay length increased when the recording medium was placed near the metallic plate. We believe this was because image charges were induced in the recording medium, and the intensity of the optical near field between the head and the recording medium became greater because of the interaction of the charges induced at both the apex of the metallic plate and the recording medium.

4.6 Near-Field Optical Head Using Two Metallic Plates

When a wedge-shaped metallic plate is used, the spot size can be further decreased by reducing the apex radius or the spacing between the head and the recording medium. An alternative way to decrease the spot size is to use two metallic plates (Fig. 29). When the two metallic plates are illuminated by light, an optical near field is generated between the two apices because of the interaction of charges concentrated at these apices. This mechanism is similar to that of the bow-tie antenna used at microwave frequencies. However, at optical frequencies, the intensity of the optical near field can be enhanced when the plasmon is excited in the metallic plates by optimizing the material and the size of the metallic plate.

Figure 30 shows the intensity distribution of the optical near field generated near the two metallic plates. For this calculation, we assumed that the shape of the metallic plates was a sector, the metal was silver, the gap width

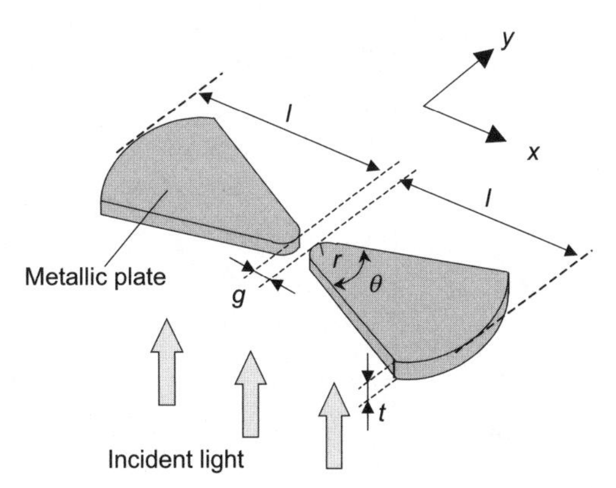

Fig. 29. Schematic of the double metallic plate

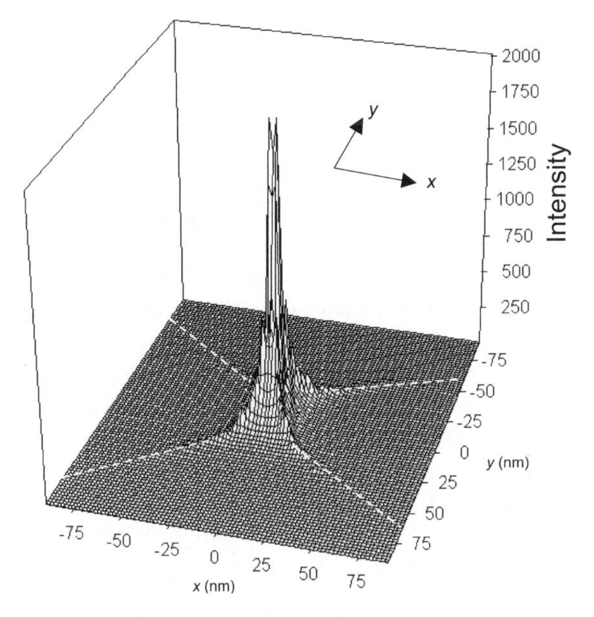

Fig. 30. Intensity distribution of the optical near field generated near the double metallic plate. It was assumed that the metallic plates had a sector shape, the metal was silver, the gap width g was 5 nm, the apex radius r was 20 nm, the length l was 200 nm, the thickness t was 30 nm, the apex angle θ was 60°, and the wavelength was 830 nm. The distribution was calculated on a plane 2 nm from the metallic plates

g was 5 nm, the radius r of each apex was 20 nm, the length l was 200 nm, the thickness t was 30 nm, the apex angle θ was 60°, and the wavelength was 830 nm. The distribution was calculated on a plane 2 nm from the metallic plates. As shown, a strong optical near field was generated at the gap. The peak intensity was about 2000 times that of the incident light. The distribu-

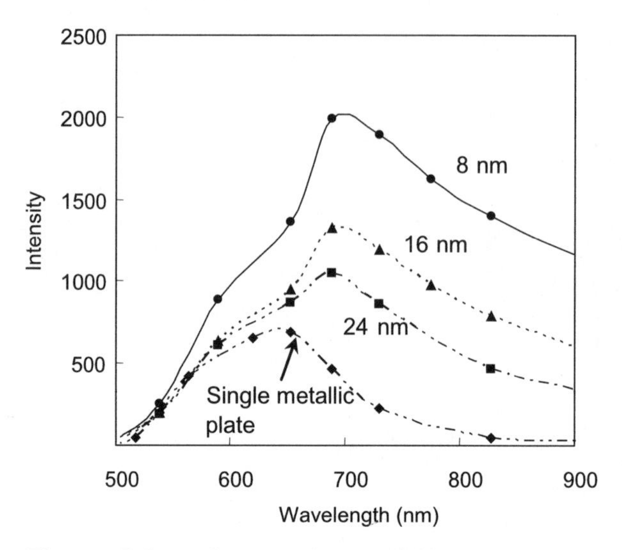

Fig. 31. Relation between the near-field intensity and the wavelength for the double metallic plates with a gap width of 24 nm, 16 nm, and 8 nm

tion measured at half-maximum was 5 nm by 5 nm, which was the same as the gap width.

Figure 31 shows the intensity of the optical near field as a function of the incident light wavelength when the gap width of two metallic plates was varied from 24 nm to 8 nm. For this calculation, we assumed that the metal was silver, the length l was 100 nm, the radius r of each apex was 20 nm, the thickness t was 30 nm, the apex angle θ was 60°. As shown, the intensity of the optical near field increased when the plasmon was excited in the metallic plates, as was the case with a single metallic plate. The resonance wavelength shifted toward a longer wavelength, and the intensity rose as the gap width became smaller because of the interaction of plasmons generated on both sides of the metallic plates.

4.7 Fabrication of a Near-Field Optical Head with a Wedge-Shaped Metallic Plate

As described in Sect. 4.1, a near-field optical head with a metallic plate can be easily fabricated by using electron beam lithography. Figure 32 shows an example of fabrication process. In this process, the head is fabricated using a sacrificial substrate to embed the metallic plate in the slider. First, an array of metallic plates is fabricated, using electron beam lithography, on a sacrificial substrate coated with a sacrificial film (a thin metallic film) in steps (a) to (c). Secondly, a photoresist pattern is formed on the sacrificial substrate to make pads near the metallic plates, and a dielectric film (Al_2O_3) is formed on the substrate by sputtering. Thirdly, a glass substrate was glued to the

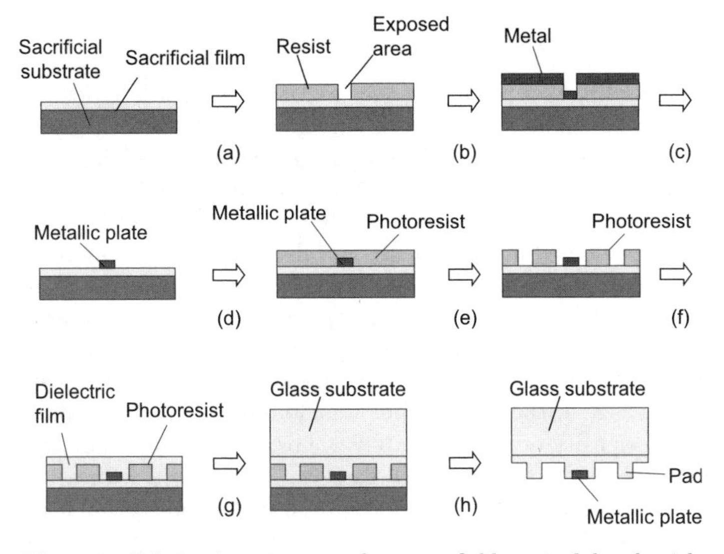

Fig. 32. Fabrication process of a near-field optical head with a wedge-shaped metallic plate: (**a**) Coating resist and exposure with a electron beam exposure machine, (**b**) coating metal, (**c**) removing resist, (**d**) coating photoresist, (**e**) exposure, (**f**) coating dielectric film, (**g**) gluing a glass substrate onto the structure, and (**h**) removing the sacrificial substrate

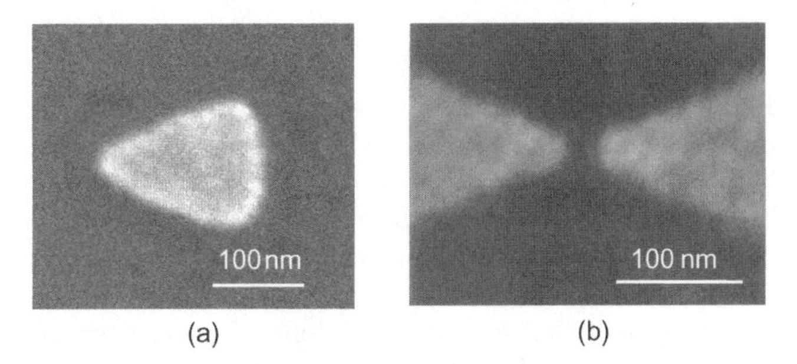

Fig. 33. SEM image of the metallic plates: (**a**) one metallic plate, (**b**) two metallic plates

top of the structure, and then the photoresist and the sacrificial substrate are removed. Figure 33 shows a scanning electron microscope image of metallic plates fabricated by the method described above. In this fabrication, gold was used as the metal, and the thickness was 30 nm. As shown in this figure, the radius of the apex was about 10 nm, and the gap width was 20 nm for the two metallic plates. Based on our simulation results, a recording density of about 1 Tb/in^2 should be achievable by using the fabricated near-field optical head.

5 Summary

Near-field optical recording is a promising way to realize a recording density of over 1 Tb/in^2. In this chapter, we focused on the near-field optical head, which is a key device for near-field optical recording. First, we explained the technical issues regarding the near-field optical head and introduced some solutions to these issues. We focused on a highly efficient near-field optical head that uses a wedge-shaped metallic plate, and described its optical properties based on a simulation using a finite-difference time-domain method. The simulation results confirmed that a strong optical near field is generated at the apex of the metallic plate when a plasmon is excited in the metallic plate. When a TbFeCo recording medium was placed 10 nm from the near-field optical head, the size of the optical spot was 30 nm, which corresponds to an areal recording density of approximately 1 Tb/in^2. The efficiency was 20% if we assume that the incident beam was a Gaussian beam with a full width at half-maximum of 1 μm. Furthermore, we discussed an optical head using two metallic plates. We confirmed through our simulation that a highly localized optical near field was generated at the gap when the plasmon was excited in the metallic plates. The distribution was 5 nm by 5 nm when the two apices were separated by 5 nm. These results indicate that the recording-density limit now being encountered in conventional recording devices can be overcome with this type of near-field optical head.

To realize such advanced recording devices, development of a recording medium suitable for near-field optical recording is also important. For instance, in the case of magnetic recording medium, a recording medium with a small grain size and thin cover layer will be needed, and tuning of the magnetic and thermal characteristics will be necessary. High-precision tracking servo technology will also be required. For example, a track pitch of less than 40 nm is required for a recording density of more than 1 Tb/in^2; that of the current Blu-ray disk is 320 nm. We believe that these technologies can be realized on the basis of technology that has been developed for conventional optical or magnetic recording devices. However, various technical challenges will have to be overcome to realize such technology because the required size and precision are less than 1/10 those of today's conventional technology.

Acknowledgements

The author would like to thank H. Sukeda, T. Shimano, M. Kiguchi, H. Saga, T. Shintani, Y. Anzai and K. Ishikawa of the Research and Development Group, Hitachi Ltd. for their helpful discussions. This work was supported by the "Terabyte optical storage technology" project, which the Optoelectronic Industry and Technology Development Association contracted with The Ministry of Economy, Trade, and Industry of Japan (METI) in 2002 based on funds provided by METI.

References

1. *Near-field Nano/Atom Optics and Technology*, ed. by M. Ohtsu (Springer, Tokyo 1998)
2. S.M. Mansfield, G. S. Kino: Appl. Phys. Lett. **57**, 2615 (1990)
3. B.D. Terris, H.J. Mamin, D. Rugar, W.R. Studenmund, G.S. Kino: Appl. Phys. Lett. **65**, 388 (1994)
4. M. Shinoda, K. Saito, T. Ishimoto, T. Kondo, A. Nakaoki, M. Furuki, M. Takeda, M. Yamamoto: In: *Technical Digest of Optical Data Storage 2003* (Canada 2003) 120
5. E. Betzig, J.K. Trautman, R. Wolfe, E.M. Gyorgy, P.L. Finn, M.H. Kryder, C.-H. Chang: Appl. Phys. Lett. **61**, 142 (1992)
6. S. Hosaka, T. Shintani, M. Miyamoto, A. Kikukawa, A. Hirtsune, M. Terao, M. Yoshida, K. Fujita, S. Krammer: J. Appl. Phys. **79**, 8082 (1996)
7. S. Jiang, J. Ichihashi, H. Monobe, M. Fujihira, M. Ohtsu: Opt. Commun. **106**, 173 (1994)
8. H. Saga, H. Nemoto, H. Sukeda, M. Takahashi: Jpn. J. Appl. Phys. **38**, Part 1, 1839 (1999)
9. H. Sukeda, H. Saga, H. Nemoto, Y. Ito, C. Haginoya: IEEE Trans. Magn. **37**, 1234 (2001)
10. M. Vasz-Iravani, R. Toledo-Crow, Y. Chen: J. Vac. Sci. Technol. A **11**, 742 (1993)
11. E. Betzig, P.L. Finn, J.S. Weiner: Appl. Phys. Lett. **60**, 2484 (1995)
12. M.B. Lee, M. Kourogi, T. Yatsui, K. Tsutsui, N. Atoda, M. Ohtsu: Appl. Opt. **38**, 3566 (1999)
13. F. Isshiki, K. Ito, K. Etoh, S. Hosaka: Appl. Phys. Lett. **76**, 804 (2000)
14. T. Yatsui, M. Kourogi, K. Tsutsui, J. Takahashi, M. Ohtsu: Opt. Lett. **25**, 1279 (2000)
15. A. Partovi, D. Peale, M. Wutting, C.A. Murray, G. Zydzik, L. Hopkins, K. Baldwin. W.S. Hobson, J. Wynn, J. Lopata, L. Dhar, R. Chichester, J.H.-J. Yeh: Appl. Phys. Lett. **75**, 1515 (1999)
16. K. Goto: Jpn. J. Appl. Phys. **37**, Part 1, 2274 (1998)
17. K. Kato, S. Ichihara, M. Oumi, H. Maeda, T. Niwa, Y. Mitsuoka, K. Nakajima, T. Ohkubo, K. Itao: Jpn. J. Appl. Phys. **42**, Part 1, 5102 (2003)
18. T.W. McDaniel, W.A. Challener: In: *Technical Digest of MORIS'02* (France 2002) 117
19. T. Saiki, S. Mononobe, M. Ohtsu, N. Saito, J. Kusano: Appl. Phys. Lett. **68**, 2612 (1996)
20. T. Yatsui, M. Kourogi, M. Ohtsu: Appl. Phys. Lett. **73**, 2090 (1998)
21. U.C. Fischer et al.: In: *Near-Field Optics* ed. by D.W. Pohl, D. Courjon (Kluwer, Academic, Dordrecht 1993) 255
22. J. Koglin, U.C. Fischer, H. Fuchs: Phys. Rev. B **55**, 7977 (1997)
23. H. Raether: *Surface Plasmon on Smooth and Rough Surface and on Grating* (Springer-Verlag, Berlin 1988)
24. T. Thio, K.M. Pellerin, R.A. Linke, H.J. Lezec, T.W. Ebbesen: Opt. Lett. **26**, 1972 (2001)
25. U.C. Fischer, D.W. Pohl: Phys. Rev. Lett. **62**, 458 (1989)
26. T. Matsumoto, T. Ichimura, T. Yatsui, M. Kourogi, T. Saiki, M. Ohtsu: Opt. Rev. **5**, 369 (1998)

27. U.C. Fischer: Ultramicroscopy **42**, 393 (1992)
28. J. Takahara, S. Yamagishi, H. Taki, A. Morimoto, T. Kobayashi: Opt. Lett. **22**, 475 (1997)
29. R. Grober, R. Schoelkopf, D. Prober: Appl. Phys. Lett. **70**, 1354 (1997)
30. T. Matsumoto, T. Shimano, S. Hosaka: In: *Technical digest of 6th international conference on near field optics and related techniques* (Netherlands 2000) 55
31. T. Matsumoto, T. Shimano, H. Saga, H. Sukeda, M. Kiguchi: J. Appl. Phys. **95**, 3901 (2004)
32. J.B. Judkins, R.W. Ziolkwski: J. Opt. Soc. Am. A **12**, 1974 (1995)
33. H. Furukawa, S. Kawata: Opt. Commun. **132**, 170 (1996)
34. *Handbook of Optical Constants of Solid*, ed. by E.D. Palik (Academic Press, Orlando 1985)
35. A. Wokaun, J.P. Gordon, P.F. Liao: Phys. Rev. Lett, **48**, 957 (1982)

Nano-Optical Media for Ultrahigh-Density Storage

K. Naito, H. Hieda, T. Ishino, K. Tanaka, M. Sakurai, Y. Kamata, S. Morita, A. Kikitsu, and K. Asakawa

1 Introduction

The data storage density is advancing because of the requirement of information data growth as shown in Fig. 1. A 1 Tb/inch^2 density will be required around 2010. Physical limits, however, are predicted both for magnetic recording and for optical means. To overcome the optical limit (diffraction limit), near-field optics is required [1].

The magnetic recording density is currently increasing at rates of up to 100 % per year. To obtain the higher densities, grain sizes of the conventional continuous magnetic films should be reduced to maintain a necessary signal-to-noise ratio. The small grain sizes, however, reduce the thermal stability of the magnetization of each bit. This, referred to as superparamagnetism, can be overcome by increasing KuV/kT, where Ku is an energy barrier to reversal per grain volume (anisotropic energy), V is a volume per grain, k is the Boltzmann constant, and T is temperature. High Ku materials such

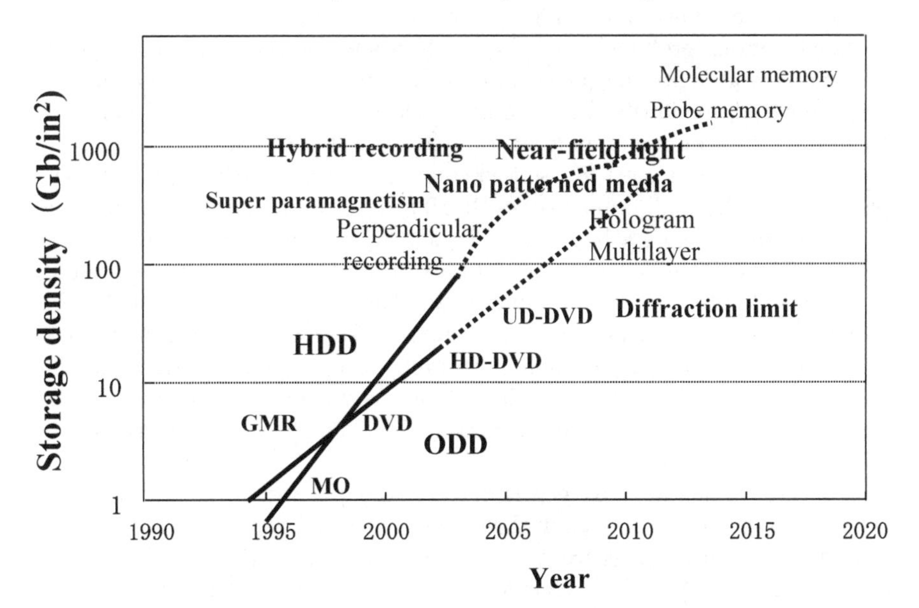

Fig. 1. Roadmap of storage technologies

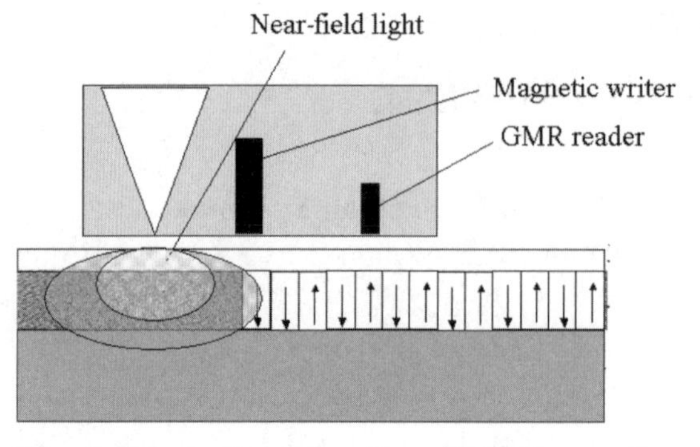

Fig. 2. Schematic explanation of hybrid recording

as FePt are preferable to overcome the thermal fluctuation. Higher magnetic fields are required to write the signals on the high-Ku media. Magnetic fields generated by the thin-film magnetic head have almost reached their limits. Hybrid recording, consisting of near-field light and a magnetic field, is one of the promising candidates to solve the paradox as shown in Fig. 2.

Magnetic patterned media, which consist of islands of magnetic material surrounded by a non magnetic matrix as shown in Fig. 3, increase an effective V resulting in bit thermal stability [2]. Magnetic patterned media can decrease media transition noise, track edge noise and nonlinear bit shift effects. Additionally, precise servo information can be embedded in the patterned media. This is very advantageous for narrower track widths of future HDDs [3]. Ultrahigh density beyond 1 Tb/inch^2 should be obtained by the combination of hybrid recording and the patterned media.

Some serious problems must be resolved such as the high cost of nanopatterning and write synchronization [4]. Magnetic patterned media have been prepared previously, for example, by electron-beam or focused-ion-beam patterning [5], ion-beam modification [6], interferometric lithography [7], nanoimprint lithography [8], self-assembling template lithography [9] and plating in anodized aluminum pores [10]. The latter four methods can produce patterned media with a large area. To our knowledge, however, there is no example of preparing circumferential patterned media necessary for HDDs.

In this chapter, we show circumferential magnetic patterned media, prepared by an artificially assisted (or aligned) self-assembling (AASA) method [11], which includes simple nanopatterning using a nanoimprint and fine nanopatterning using self-assembling diblock polymers [12]. The fine nanopatterns created by the self-assembling method can be aligned in a large area by the artificial nanoimprint.

We have aimed at a novel write/read principle media on the nm scale: organic-dye-patterned media. In the previous near-field optical storage demonstrations, conventional recording media such as a magnetic-optical recording (MO) medium [13], a phase-change (PC) medium [14], or photochromic materials [15, 16] have been used. Using an MO medium and a PC medium,

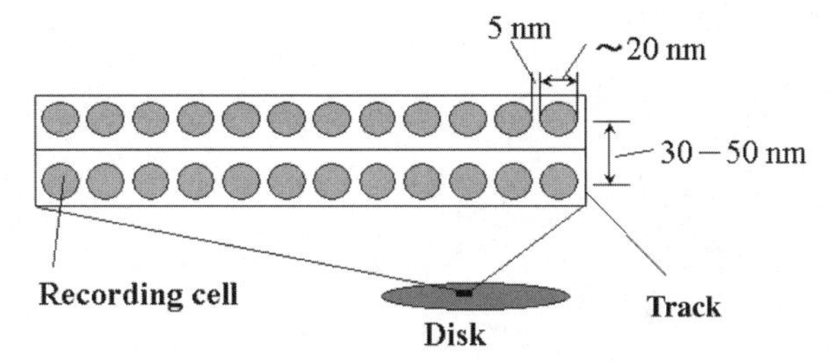

Fig. 3. Schematic explanation of patterned media

signal contrasts are 1% and 10%, respectively. In future near-field storage devices, signal contrast will be insufficient to achieve practical read-out speeds. As for photochromic materials, although better signal contrast is expected, read out without destroying the recorded information has been a major issue.

We showed that amorphous organic dye molecules formed droplet-like nm-scale structures and were aligned by the AASA method [17]. It is possible to inject and confine charges into a single dot using AFM [18, 19]. Fluorescence changes in dye molecules associated with charge injection into their thin films was observed [20]. Fluorescence measurement has such a high signal-to-noise ratio that it can be observed even in a single molecule. Therefore, read out using fluorescence is expected to achieve high-speed scanning. Furthermore, electrical writing, namely charge confinement, will prevent record rupture at read out.

2 Magnetic Patterned Media

2.1 Preparation

Figure 4 shows a schematic explanation of the preparation method. The self-assembly method has attracted much attention because of its ability to produce ordered fine nanopatterns with a large area. We adopted diblock copoly-

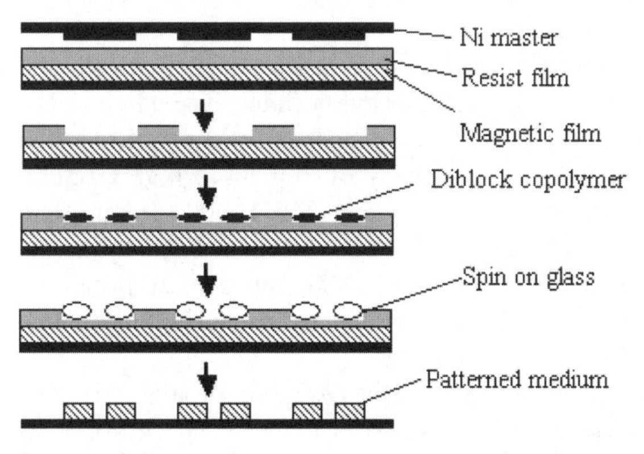

Fig. 4. Scheme of the preparation method of patterned media

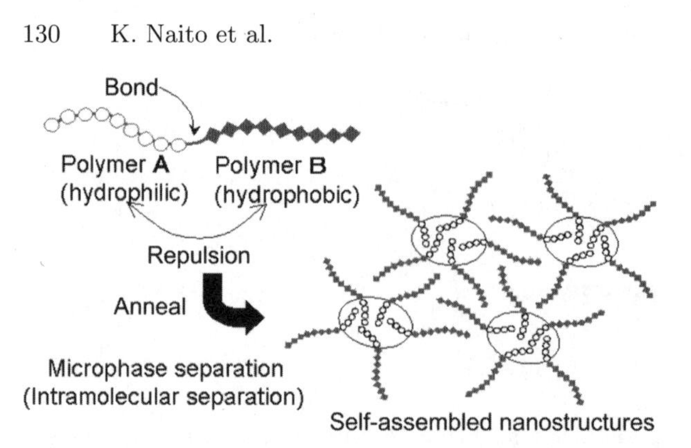

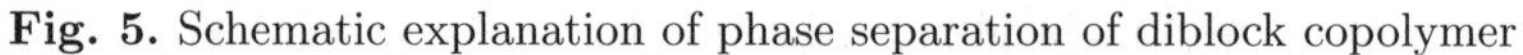

Fig. 5. Schematic explanation of phase separation of diblock copolymer

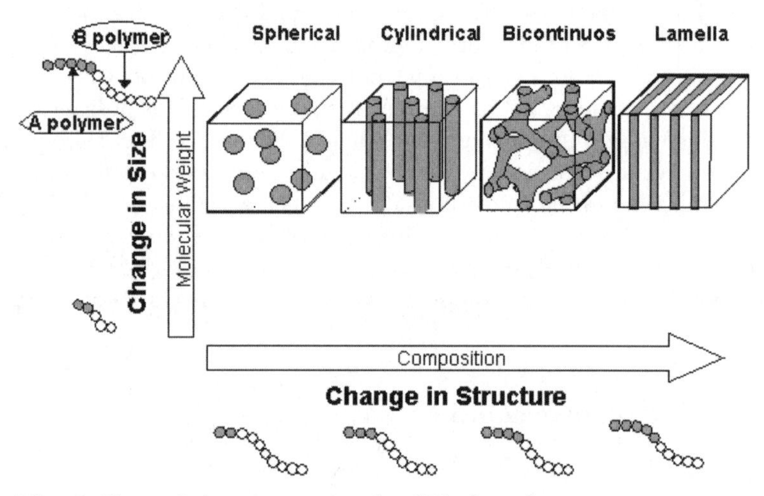

Fig. 6. Type of phase separation for diblock coplymer

mers as the self-assembly system. Figure 5 shows the explanation of phase-separation of diblock coplymers. There are several types of phase-separated states as shown in Fig. 6.

The diblock copolymer molecule used here consists of a hydrophobic polystyrene chain and a hydrophilic polymethylmethacrylate chain [21]. These chains are not mixed resulting in a phase-separation state. The dot diameter and the dot pitch are determined by the molecular weights. For the diblock copolymer [polymethylmethacrylate (Mw 41 500)-polystyrene (Mw 17 200)], 40 nm diameter and 80 nm pitch were obtained. Figure 7 shows the dot patterns of the diblock copolymers on a flat surface. A polycrystalline structure was observed for the phase-separation state with polymethylmethacrylate dots in a polystyrene surrounding.

Figure 8 shows the consept of the artificially assisted (or aligned) self-assembly (AASA) method. Phase separation takes place in the grooves to produce aligned dot patterns.

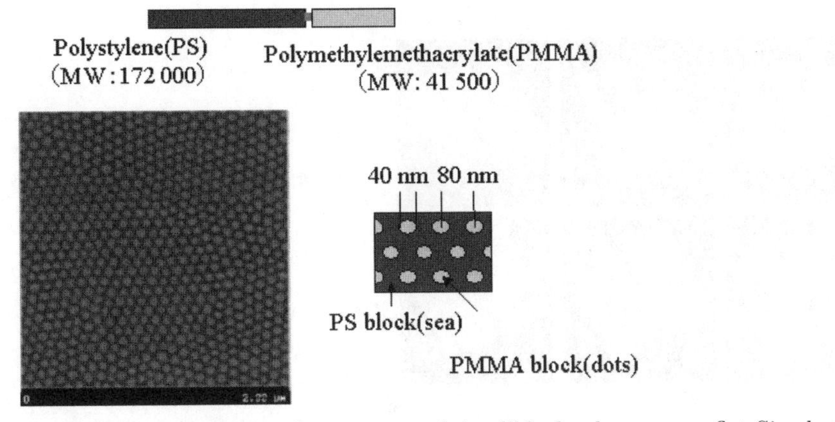

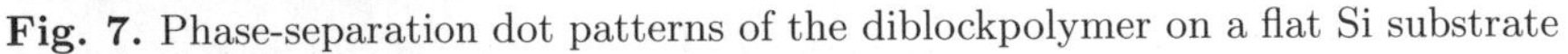

Fig. 7. Phase-separation dot patterns of the diblockpolymer on a flat Si substrate

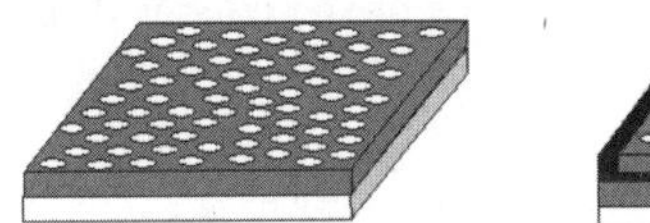

Fig. 8. Concept of AASA method

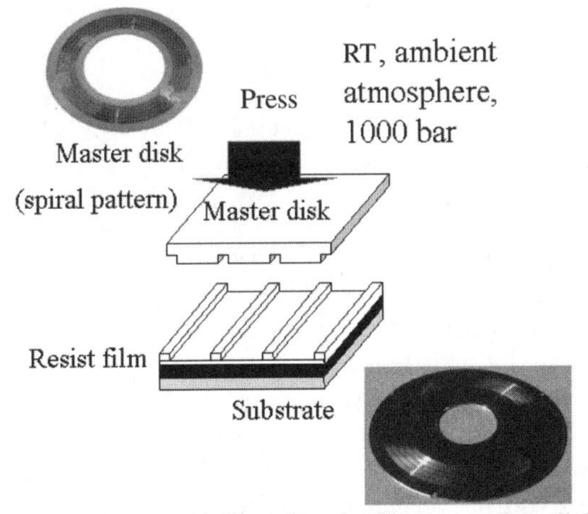

Fig. 9. The whole image of the Ni master

A Ni master disk possessing spiral patterns with 60-, 100-, 150-, 200- and 250-nm width lands and a 400-nm width groove (110-nm depth) was pressed to a resist film on a CoCrPt film to transfer the spiral patterns at room temperature and at a pressure of 1000 bar. The master disk was prepared by the photoprocess as shown in Fig. 9.

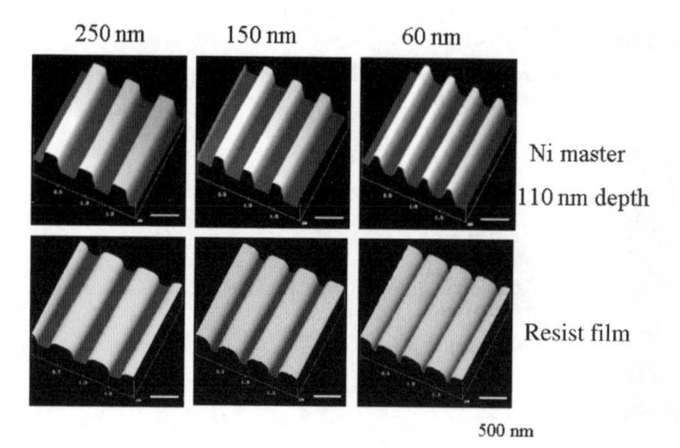

Fig. 10. AFM images of spiral patterns of the Ni master (*upper images*) and of the resist film (*lower images*). The lengths are the land widths of the master disk

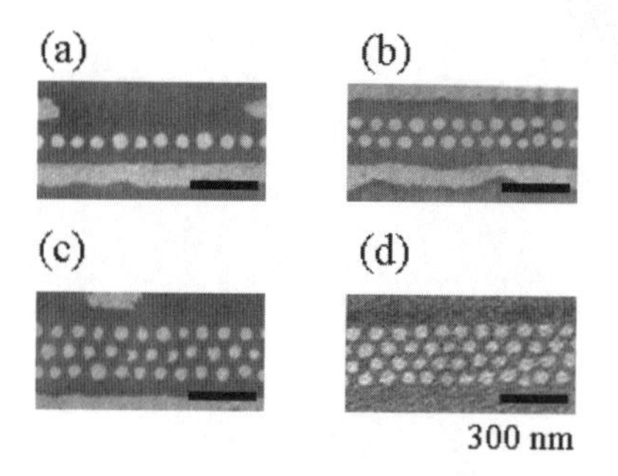

Fig. 11. Phase-separation dot patterns of the diblock copolymers in grooves with different widths: (**a**) 60 nm, (**b**) 150 nm, (**c**) 200 nm, (**d**) 250 nm

The AFM images of spiral patterns transferred to the resist film are shown in Fig. 10. The land-groove spiral patterns were precisely transferred.

A diblock copolymer solution was cast into the obtained grooves, and then annealed to prepare self-assembling dot structures aligned along the grooves. Figure 11 indicates dot structures in grooves with different widths. One to four dot lines could be obtained by changing the groove width. The groove width should be controlled to obtain the regularly aligned dot lines along the grooves. In the figure, the groove widths were considered rather suitable for the two and four dot lines. Figure 12 shows the phase-separation dot patterns of another diblock copolymer in grooves. The dot size was 15 nm and the dot pitch was 35 nm. This small dots were obtained by using the low molecular weight diblock copolymer.

The PMMA dots were selectively removed by the oxygen plasma treatment [12]. The resulting holes were filled by spin-on-glass (SOG). The lower magnetic films were patterned by ion milling using the SOG dots as a mask.

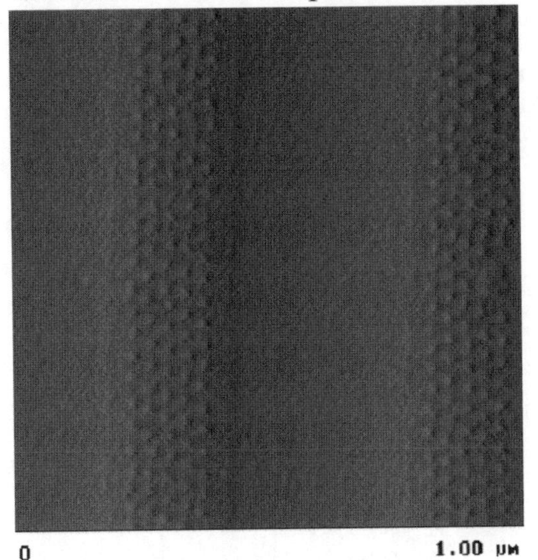

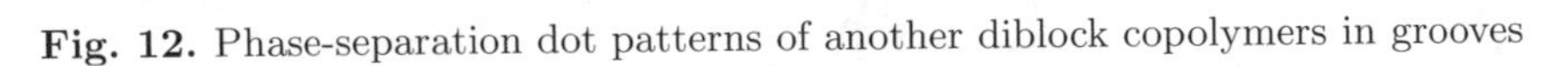
Fig. 12. Phase-separation dot patterns of another diblock copolymers in grooves

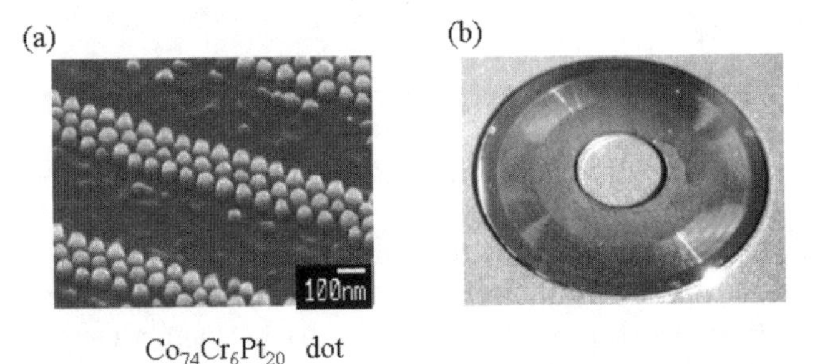

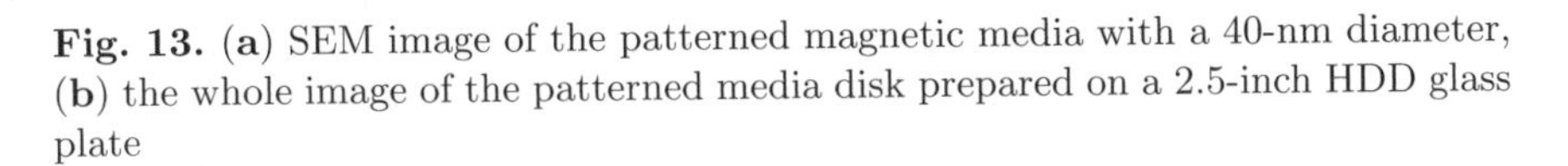
Fig. 13. (**a**) SEM image of the patterned magnetic media with a 40-nm diameter, (**b**) the whole image of the patterned media disk prepared on a 2.5-inch HDD glass plate

Figure 13a shows the SEM image of the patterned magnetic media with 40 nm diameter. There are fluctuations in size and in position of the magnetic dots. The irregularities are probably caused by the diblock copolymer dot disorder and by the nanopatterning process fluctuation. The whole image of the patterned media disk prepared on a 2.5-inch diameter HDD glass plate is shown in Fig. 13b. Interference colors based on the wide groove lines were observed. Figure 14 shows the TEM image of the patterned dot with about 40 nm diameter. This image indicates that the magnetic dot is completely separated.

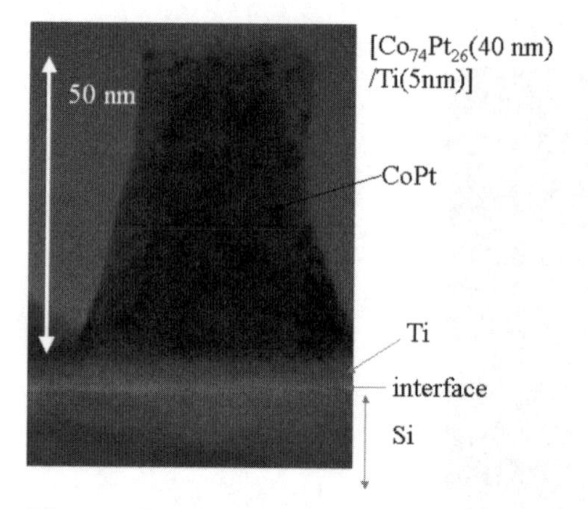

Fig. 14. TEM image of the patterned CoPt dot

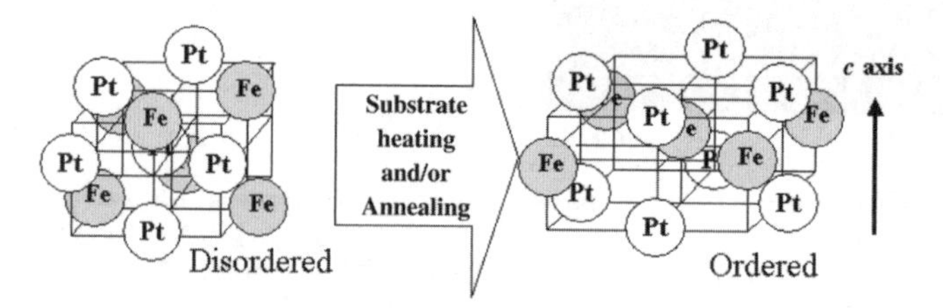

Fig. 15. Schematic explanation of ordering of FePt alloy

FePt is an ordered alloy that has a high Ku. High temperature is required to obtain the ordered structure form the sputtered film as shown in Fig. 15. We found that addition of Cu decreases the temperature [22].

Using an electron-beam mastering apparatus recently developed will enable reduction of the groove width to increase the density of magnetic cells. It may be impossible to produce circumferential dot patterns with a 40 nm diameter like those in the figure even by the present electron beam apparatus.

We used the AASA method to prepare the etching mask for the magnetic films. For cheaper preparation, we prepared the Ni master disk itself by the AASA method. The details will be shown elsewhere.

2.2 Magnetic Properties

Figure 16 shows the M–H loops of the raw continuous film (CoCrPt) and the patterned media. A coercive force and squareness ratio of the patterned film increased compared to the continuous film. The magnetization of the patterned media decreased due to the decrement of the material volume. The squareness ratio was almost unity.

Figure 17 indicates the MFM images of the patterned magnetic media after AC erase and after DC erase. Three magnetic dot lines are observed corresponding to Fig. 11c and Fig. 13a. Bright and dark dots are randomly

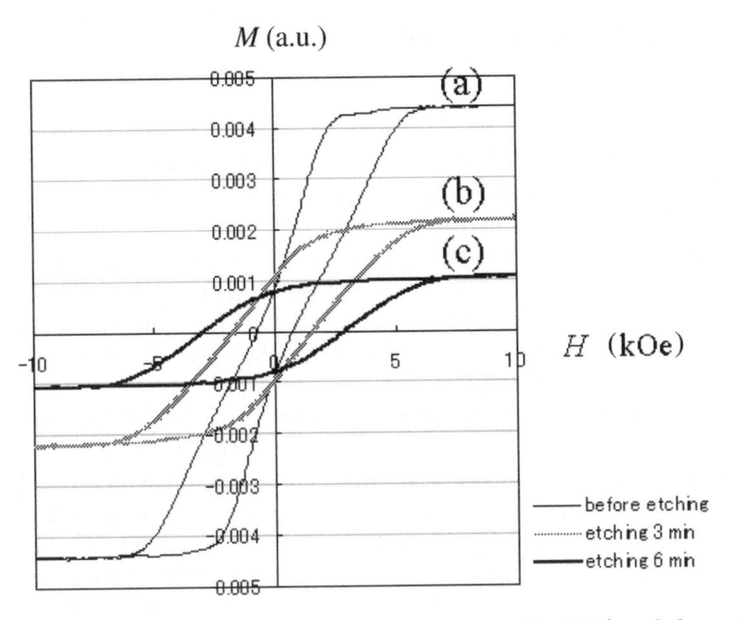

Fig. 16. M–H loops of the raw continuous film (CoCrPt) and the patterned media

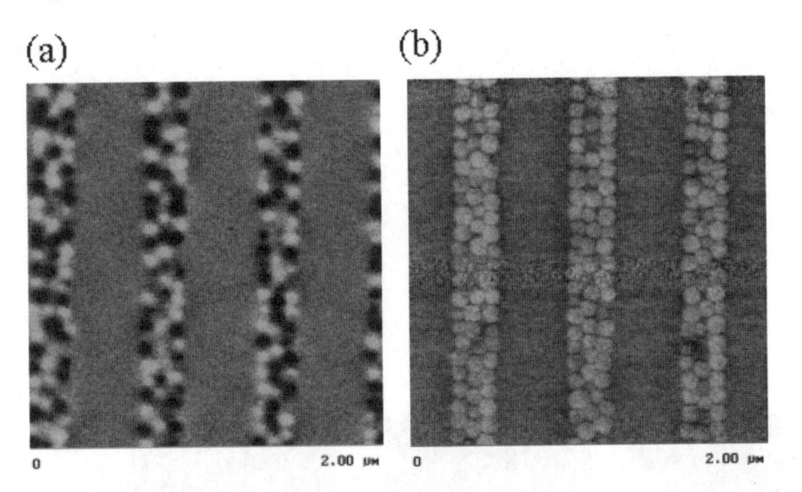

Fig. 17. MFM images of the patterned media

distributed after AC erase. The DC-erased dots are observed to be magnetized in one direction in Fig. 17b. Single magnetic domains with an almost perpendicular orientation were confirmed in each magnetic dot.

3 Organic-Dye-Patterned Media

3.1 Preparation

The organic molecule mainly used here was TTPAE or TPD, shown in Fig. 18. These molecules give amorphous films as a hole transport layer in organic EL devices [23]. The films were formed by vacuum evaporation onto thermally oxidized p-Si substrates with a resistivity of 0.001–0.01 Ω cm. The thickness of the SiO_2 layer was 20 nm. The electrodes were fabricated on the back of the Si substrates by evaporating Cr and Au. Before the evaporation of molecules, the SiO_2/Si substrates were made hydrophobic by treatment in an atmosphere of hexamethyldisilazane.

Figure 19 shows the contact-mode topographic image of the TTPAE film observed using an AFM. It is clearly seen that the film is constructed of spatially isolated droplet-like domain structures. The size and the density of domains were varied according to the amount of deposition. The diameter, height and density were varied from 20 to 300 nm, from 10 to 100 nm, and from 1011 to 109 cm^{-2}, respectively. When the amount of deposition was

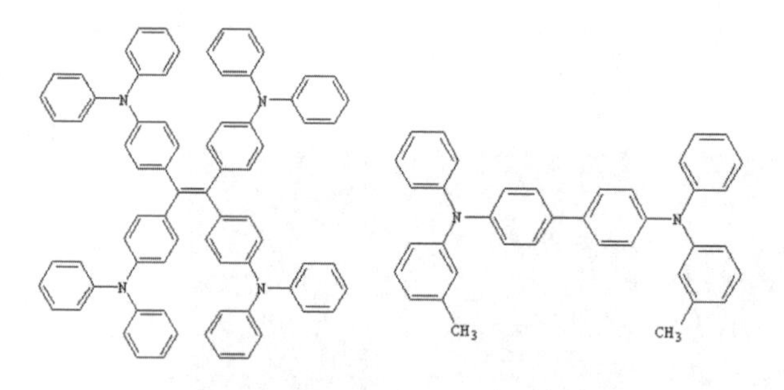

Fig. 18. Molecular structures of TTPAE and TPD

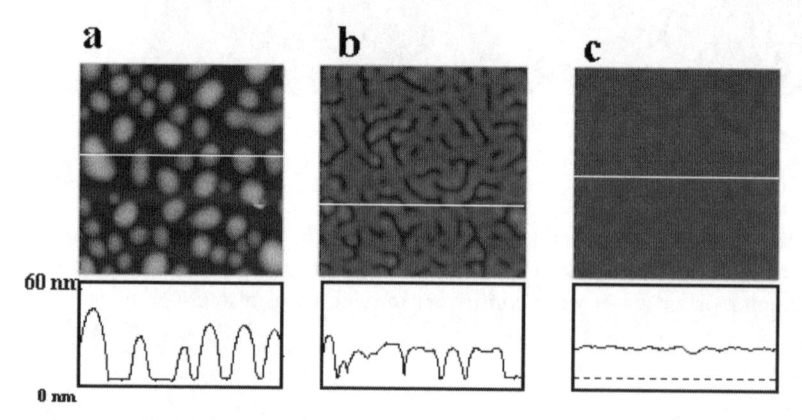

Fig. 19. AFM image of TTPAE films

increased to the film thickness of more than 40 nm, the film structure changed into a continuous layer with a flat surface. We verified that thin films of several amorphous molecules, including TPD, also showed similar domain structures.

The observed uniform droplet-like domain structures suggest that the film growth is based on the Volmer–Weber mechanism. In this mechanism, growth characteristics are known to depend strongly on the surface energies of the film and of the substrate. When the films were formed onto hydrophilic SiO_2/Si substrates, it was found that the domain size was much larger than films on the hydrophobic substrates.

The droplet size and alignment could be controlled by the AASA method. The hole arrays were prepared on polymethylmethacrylate film by using AFM. The pitch was 30 nm. TPD molecules were evaporated onto the arrays. TPD amorphous dots were selectively formed on the holes with the same size as shown in Fig. 20. Using a nanoimprint method should produce the large-area hole arrays to obtain the large-area organic-dye-patterned media as shown in Fig. 21.

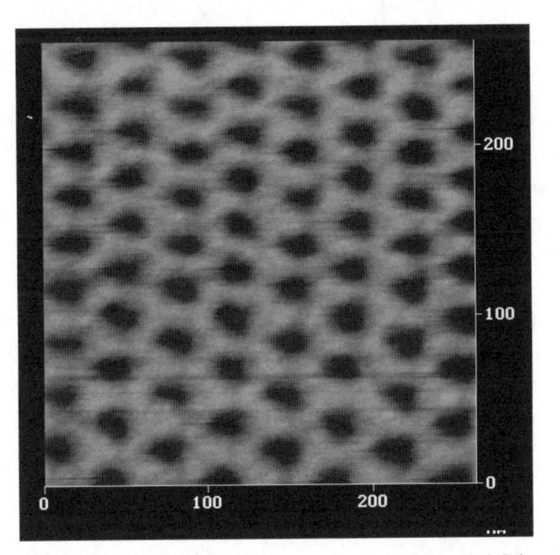

Fig. 20. TPD dot patterns prepared by AASA method

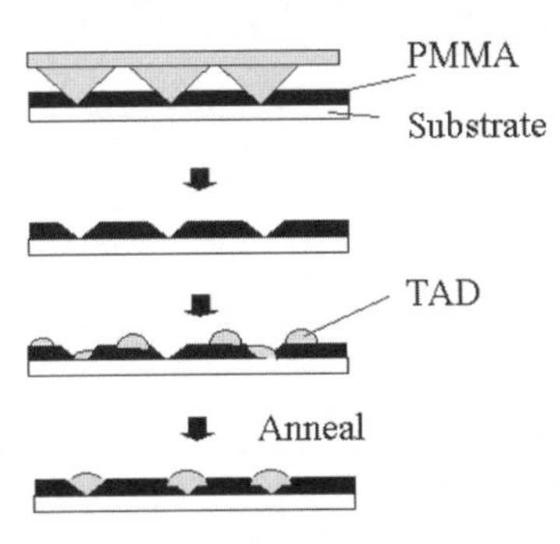

Fig. 21. Schematic explanation of TPD dot alignment

3.2 Electrical and Optical Measurements

Figure 22 shows a schematic diagram of the structure of the device used in this chapter. To apply voltage to the organic films, Au films of 10 nm thickness were fabricated on the substrates following the deposition of organic films. Au films of 10 nm were selected as the thinnest providing good electrical contact and sufficient fluorescence intensity from the underlying organic films.

Scanning of the applied voltage started at 0 V. Fluorescence measurements were first obtained in the positive-voltage regions, and then in the negative regions. The voltage scan speed was 0.02 Hz. All fluorescence measurements were carried out in air and at room temperature.

In order to investigate the electrical properties of TTPAE films, displacement-current measurements were performed. Since details of displacement-current measurements were described in [24], only distinctive features are shown here. Figure 23 shows a schematic diagram of the displacement-current measurement. The measured current is expressed by the capacitance and scan speed of the applied voltage. In the simplest case, the organic film behaves as a dielectric and the measured current reflects the capacitance of the entire dielectric film, that is, organic and SiO_2 (C_{SiO_2+Org}). As for the donor or acceptor molecules, carrier injection or ejection through metal/org interfaces

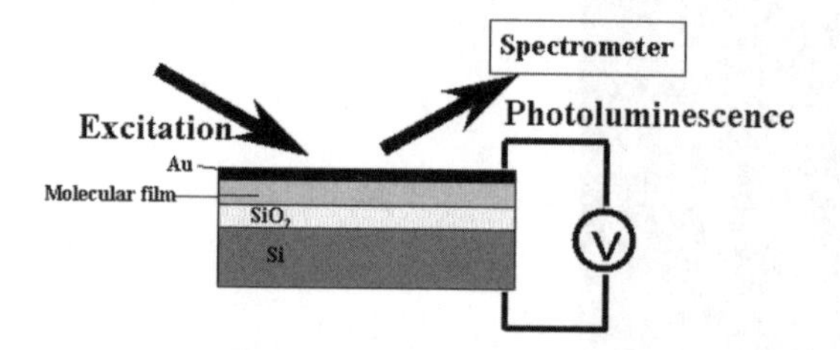

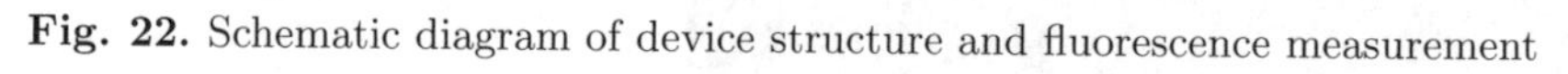
Fig. 22. Schematic diagram of device structure and fluorescence measurement

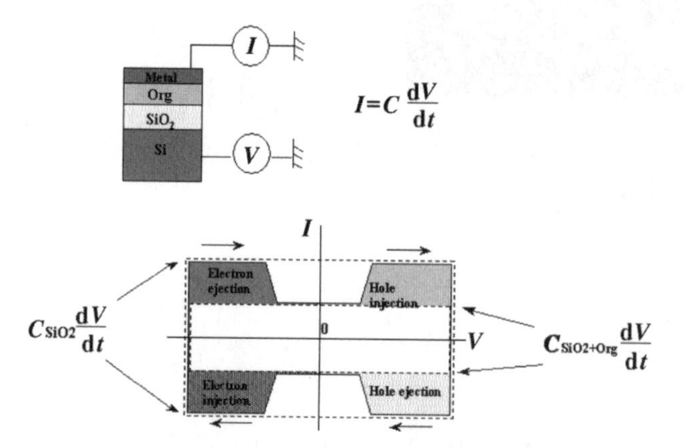

Fig. 23. Schematic diagram of displacement-current measurement

occur. If carrier injection through the metal/org interface occurs, the measured current will reflect the capacitance of the SiO_2 (C_{SiO_2}) film. Therefore, the displacement current is sensitive to carrier injection into the organic film.

Figure 24 shows the applied-voltage dependence of fluorescence in the TTPAE film. These data were obtained in the first scan of voltage for positive and negative voltages, respectively. As shown later, large hysteresis effects were observed in fluorescence quenching. As shown in Fig. 24, the fluorescence intensity decreases as applied voltage V increases at positive voltages. On the other hand, at negative voltages, no variations with V are observed. Changes in the fluorescence spectrum and excitation spectrum are shown in Fig. 25. It

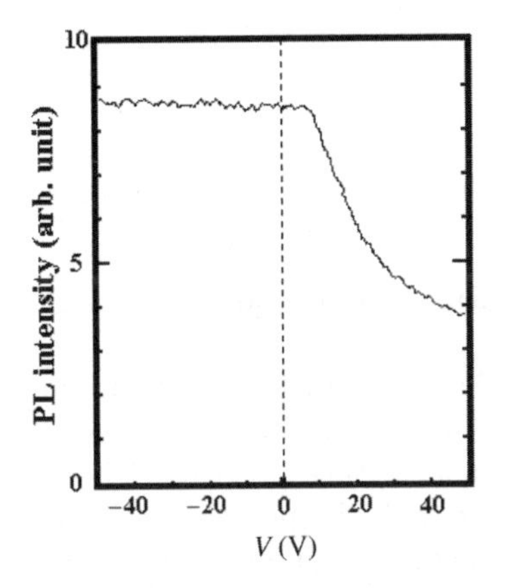

Fig. 24. Applied-voltage dependence of fluorescence (PL) of Au/TTPAE/SiO_2/Si with TTPAE thickness of 15 nm and SiO_2 thickness of 100 nm. Excitation wavelength: 360 nm. Fluorescence wavelength: 540 nm

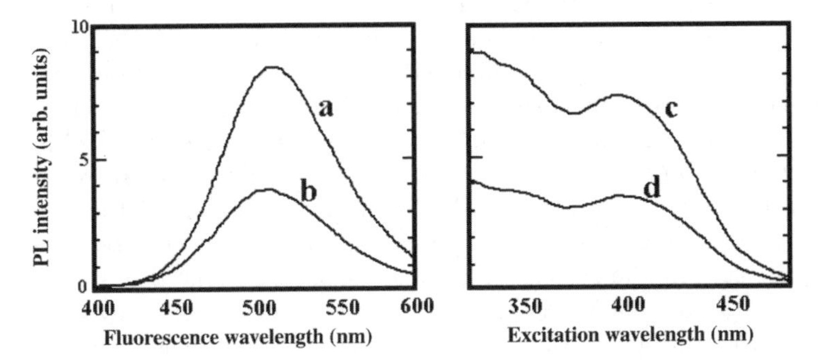

Fig. 25. Spectra of Au/TTPAE/SiO_2/Si with TTPAE thickness of 15 nm and SiO_2 thickness of 100 nm: (**a**) fluorescence spectra with bias voltage of 0 V, excitation wavelength: 360 nm; (**b**) fluorescence spectra with bias voltage of +50 V, excitation wavelength: 360 nm; (**c**) excitation spectra with bias voltage of 0 V, fluorescence wavelength of 540 nm; (**d**) excitation spectra with bias voltage of +50 V, fluorescence wavelength of 540 nm

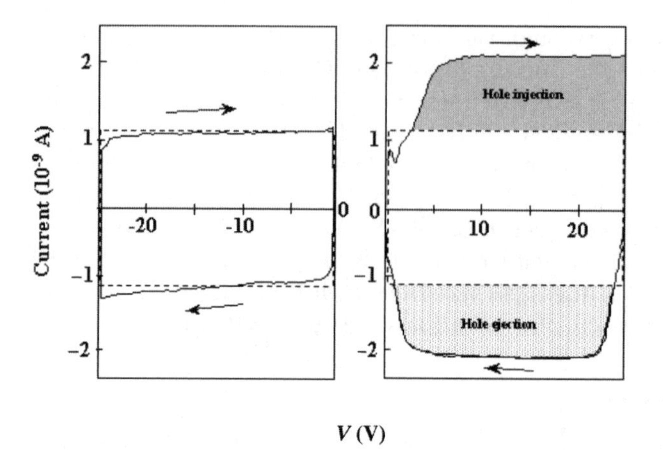

Fig. 26. Displacement-current diagram of Au/TTPAE/SiO_2/Si with TTPAE thickness of 50 nm and SiO_2 thickness of 50 nm under white-light irradiation

was found that the observed fluorescence quenching did not follow any shift of spectral peaks.

Figure 26 shows the displacement-current characteristic of a TTPAE film under visible light irradiation. At positive bias voltages, hole injection and ejection were clearly observed, and at negative bias voltages, no carrier injection was observed. These findings are consistent with the fact that TTPAE is an electron-donor material. Figure 27 shows a comparison between fluorescence measurements and displacement-current measurements under UV irradiation. Since both fluorescence variations and displacement-currents show large hysteresis especially at negative bias voltages, the results for the first cycle of voltage are shown separately from those for the second and third cycles. At positive bias voltages, fluorescence and current into the TTPAE film were observed to vary in the same manner. Both fluorescence variations and displacement currents are reproducible at positive bias voltages. On the other hand, at negative bias voltages, large hysteretic variations were observed in both measurements. On the way to -25 V in the first cycle, no fluorescence change or negative current into the TTPAE film was observed. On the way back to 0 V, a positive current and fluorescence quenching were observed at the same voltage. In the second or third cycle, two current peaks were observed on the way to -25 V and one peak was observed on the way back to 0 V. Fluorescence quenching was observed at lower voltages where current was observed on the way to -25 V, and at the same voltage as the current observed on the way back to 0 V.

Figure 28 shows the fluorescence-quenching efficiency for various thicknesses of TTPAE films. The fluorescence-quenching efficiency increased as the film thickness decreased and had a peak value at a thickness of 15 nm. In the case of films thicker than 15 nm (c), uniform film structures were

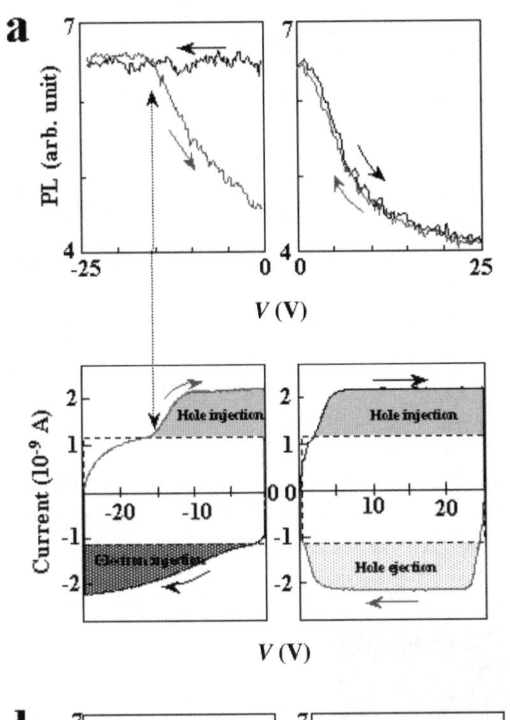

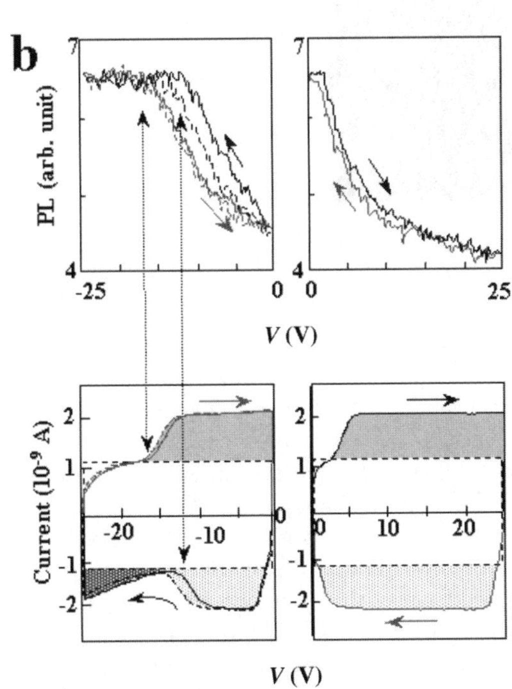

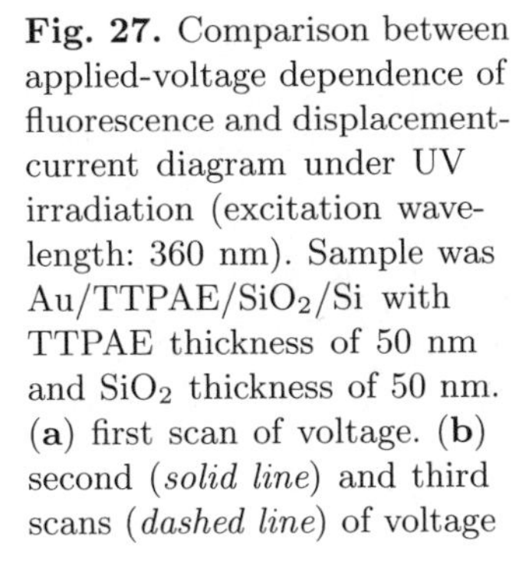

Fig. 27. Comparison between applied-voltage dependence of fluorescence and displacement-current diagram under UV irradiation (excitation wavelength: 360 nm). Sample was Au/TTPAE/SiO_2/Si with TTPAE thickness of 50 nm and SiO_2 thickness of 50 nm. (**a**) first scan of voltage. (**b**) second (*solid line*) and third scans (*dashed line*) of voltage

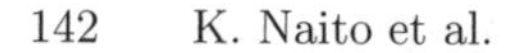

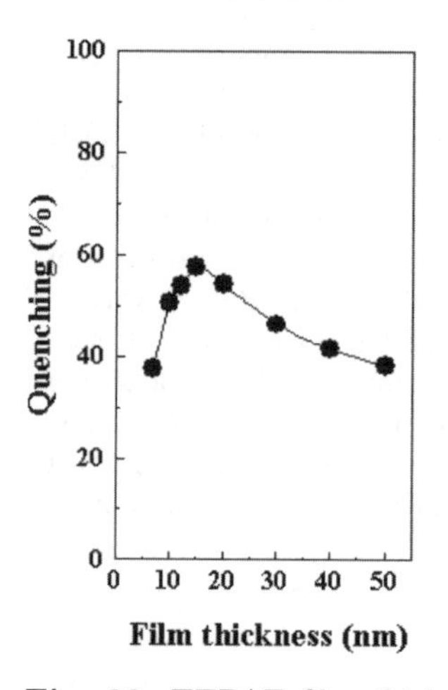

Fig. 28. TTPAE film thickness dependence of fluorescence quenching efficiency. Sample was Au/TTPAE/SiO_2/Si with SiO_2 thickness of 100 nm. Excitation wavelength: 360 nm. Fluorescence wavelength: 540 nm

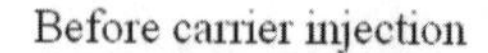

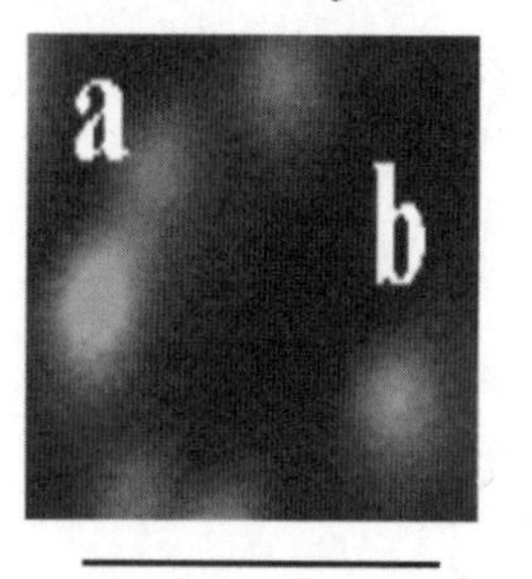

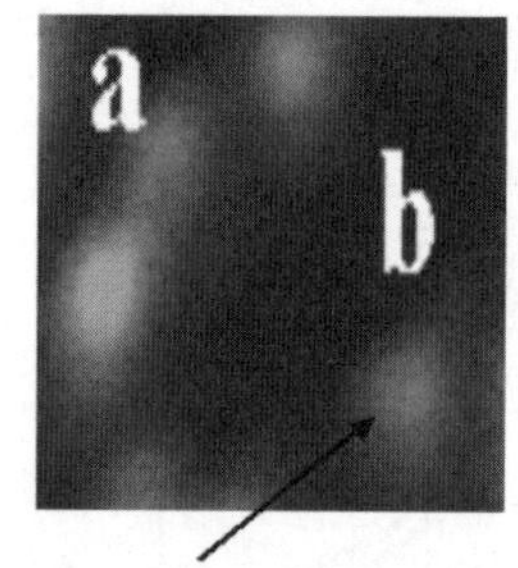

Fig. 29. NSOM images of fluorescence-intensity reduction caused by charge injection

observed. On the other hand, in the case of thinner films, separated film structures (b) or isolated dot structures (a) were observed. As for the separated structures or the dot structures, the effective thickness of the film was found to be thicker than the case for uniform films.

We directly observed fluorescence quenching for the TTPAE dots by a near-field optical microscope. Charge injection is carried out through contact electrification by a gold-coated AFM tip with a positive bias. The charges were found to be very stable over a long period. This phenomena will be the same as in a flash silicon memory, where electrons are stored in a floating gate. Injected holes can be extracted by a negatively biased tip. Holes were injected to diamine dots at 3 V. These holes were extracted at −2 V.

Therefore, the organic dot structure can be operated as a rewritable memory. However, it takes a time to detect surface potential values. Optical read-out procedures are required for high-speed scanning. Figure 29 shows that fluorescence intensity is reduced for the dot where the charge was injected.

We hypothesize that Coulomb interaction between photoexcited molecules and carriers (holes), as suggested by Deussen et al. [25] in an earlier explanation of the effect of current flow on fluorescence quenching, plays an essential role in the fluorescence quenching described here.

4 Conclusion

We have described the circumferential magnetic patterned media with a 40-nm diameter on a 2.5-inch diameter glass plate. The media were prepared by an artificially assisted self-assembling (AASA) method, which includes simple nanopatterning of land-groove spirals using a nanoimprint and self-assembling of the diblock polymer aligned in the grooves. Magnetic films were etched by ion milling using the nanodot structures of the diblock copolymer as a mask. Fluctuations of size and position of the patterned cell, observed for the present media, should be reduced for the realization of patterned media.

We have also investigated organic-dye-patterned media prepared by the AASA method. Fluorescence quenching of TTPAE dots with charge injection applying bias voltages. Displacement-current measurements show that the quenching efficiencies depend strongly on the polarities of injected charges and that quenching cannot be explained by the external-field-induced exciton dissociation. Charge injection properties show that the hole injections play an essential role in fluorescence quenching. A 60% fluorescence-quenching efficiency is achieved. This will be a novel optical storage medium with high contrast and high signal-to-noise ratio.

These nm-scale dot structures should prove to be effective in the application of near-field optics to ultrahigh-density storage media.

Acknowledgements

The authors are very grateful to T. Maeda for providing them with the samples of the magnetic films. They also acknowledge useful discussions with H. Yoda, J. Akiyama, T. Hiraoka and Y. Yanagita.

References

1. Modified by K.N. from the roadmap (Fig. 2.3.1.1) In: *Research survey on near-field light technology*, OITDA (2000)
2. S.Y. Chou, M.S. Wei, P.R. Krauss, P.B. Fisher: J. Appl. Phys. **76**, 6673 (1994)
3. R.L. White, R.M.H. New, RE.F.W. Pease: IEEE Trans. Magn. **33**, 990 (1997)

4. R.L. White: J. Magn. Magn. Mater. **209**, 1 (2000)
5. C.T. Rettner, M.E. Best, B.D. Terris: IEEE Trans. Magn. **37**, 1649 (2001)
6. C. Chappert, H. Bernas, J. Ferre, V. Kottler, J.-P. Jamet, Y. Chen, E. Cambril, T. Devolder, F. Rousseaux, V. Mathet, H. Launois: Science **280**, 1919 (1998)
7. C.A. Ross, H.I. Smith, T. Savas, M. Schattenburg, M. Farhoud, M. Hwang, M. Walsh, M.C. Abraham, R.J. Ram: J. Vac. Sci. Technol. B. **17**, 3168 (1999)
8. B. Cui, W. Wu, L. Kong, X. Sun, S.Y. Chou: L. Appl. Phys. **85**, 5534 (1999)
9. T. Thurn-Albrecht, J. Schotter, G.A. Kästle, N. Emley, T. Shibauchi, L. Krusin-Elbaum, K. Guarini, C.T. Black, M.T. Tuominen, T.P. Russell: Science **290**, 2126 (2000)
10. K. Nielsch, F. Muller, A.-P Li, U. Gosele: Adv. Mater. **12**, 582 (2000)
11. K. Naito, H. Hieda, M. Sakurai, Y. Kamata, K. Asakawa: IEEE Trans. Magn. **38**, 1949 (2002)
12. R.A. Segalman, H. Yokoyama, E.J. Kramer: Adv. Mater. **13**, 1152 (2001)
13. E. Betzig, J.K. Trautman, R. Wolfe, E.M. Gyorgy, P.L. Finn, M.H. Kryder, C.-H. Chang: Appl. Phys. Lett. **61**, 142 (1992)
14. S. Hosaka, T. Shintani, M. Miyamoto, A. Hirotsune, M. Terao, M. Yoshida, S. Honma, S. Kammer: Thin Solid Films **273**, 122 (1996)
15. S. Jiang, J. Ichihashi, H. Monobe, M. Fujihira, M. Ohtsu: Opt. Commun. **106**, 173 (1994)
16. M. Hamano, M. Irie: Rev. Laser Eng. **24**, 1045 (1996) (in Japanese)
17. H. Hieda, K. Tanaka, N. Gemma: J. Appl. Phys. **40**, 1071 (2001)
18. N. Gemma, H. Hieda, K. Tanaka, S. Egusa: Jpn. J. Appl. Phys. **34**, L859 (1995)
19. H. Hieda, K. Tanaka, N. Gemma: J. Vac. Sci. Technol. B **14**, 1234 (1996)
20. H. Hieda, K. Tanaka, K. Naito, N. Gemma: Thin Solid Films **331**, 152 (1998)
21. K. Asakawa, T. Hiraoka: Annual APS march 2000 Meeting, I33-2
22. T. Maeda T. Kai, A. Kikitsu, T. Nagase, J. Akiyama: Appl. Phys. Lett. **80**, 2147 (2002)
23. K. Naito, M. Sakurai, S. Egusa: J. Phys. Chem. A **101**, 2350 (1997)
24. S. Egusa, N. Gemma, A. Miura, K. Mizushima, M. Azuma: J. Appl. Phys. **71**, 2042 (1992)
25. M. Deussen, M. Scheidler, H. Bassler: Synth. Met. **73**, 123 (1995)

A Phenomenological Description of Optical Near Fields and Optical Properties of N Two-Level Systems Interacting with Optical Near Fields

A. Shojiguchi, K. Kobayashi, S. Sangu, K. Kitahara, and M. Ohtsu

1 Introduction

1.1 What are Optical Near Fields?

Since the insight of Synge in 1928 [1], the localization of light near a material system has been recognized as one of the unique properties of an optical near field, and for more than two decades a variety of theoretical and experimental studies have been intensively conducted [2–9]. At the beginning of this chapter, we explain the spatial localization property of light, using two examples, in order to give a physical insight into the optical near-field phenomenon.

The first example is evanescent light, which is generated on a dielectric surface, as is well known, when the incident angle θ of a plane wave of light exceeds the critical angle θ_c and total internal reflection occurs as shown in Fig. 1 [2, 10]. The penetration or attenuation depth normal to the surface (along the z axis) is of the order of the wavelength of light, which indicates the spatial localization of light along the z axis. This differs from ordinary propagating light, and much attention has been paid to its characteristics [11–27]. Note that the total internal reflection is a macroscopic phenomenon, and that it is valid only on a scale at which the refractive index n of bulk material has meaning.

Another example is illustrated in Fig. 2 [2,9]. Figure 2a shows a plane wave of light incident on a plane with a small hole of radius a that is much smaller than the wavelength λ of light. Little of the propagating far-field light passes through the hole, but light exists as an optical near field near the aperture on the other side. The localization range of the optical near field is much smaller than the wavelength of light and is determined by the aperture size a. This kind of field, which a scanning near-field optical microscope (SNOM) uses at the tip of an optical near-field probe, has been investigated in detail [28–32]. A more general situation is depicted in Fig. 2b. By irradiating a dielectric that is much smaller than the wavelength of the incident light, optical near fields coiling around the object are generated together with scattering far fields. The localization range of the optical near fields is independent of the wavelength of the incident light and is determined by the size of the object: it depends on the radius a of the dielectric sphere as shown in Fig. 2b. Such spatial localization reflecting the structure of the source of the fields (i.e.,

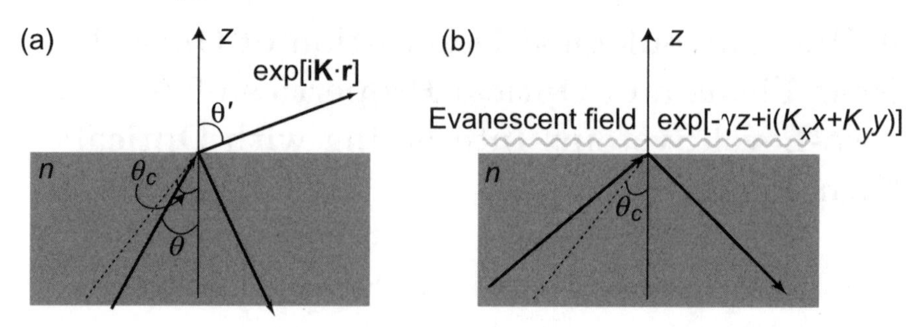

Fig. 1. (**a**) The infinite planar boundary between a dielectric surface and a vacuum. Incident light from the dielectric surface is generally reflected and refracted. (**b**) Evanescent light on the dielectric surface generated by total internal reflection of the incident light

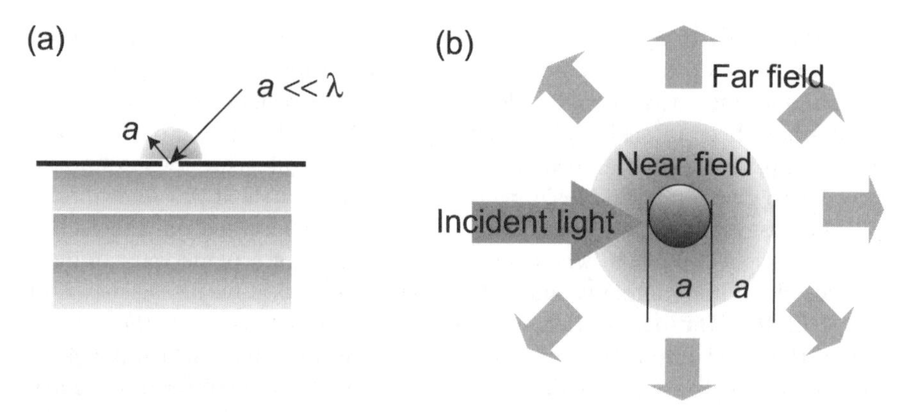

Fig. 2. Optical near fields in various boundary conditions. (**a**) An optical near field leaking from a hole that is much smaller than the wavelength of light. The leaking field attenuates as it leaves the aperture. The localization range of light is much smaller than the wavelength of light, depending on the size of the aperture. (**b**) An optical near field surrounding a small dielectric sphere and the scattered optical far field produced by an incident optical far field. The localization range of the optical near field is determined by the sphere radius and is independent of the wavelength of light

the induced polarization in matter) provides SNOM and other nanophotonic devices with spatial resolution much higher than the diffraction limit [2–4, 9]. Readers interested in a variety of SNOM applications and the state-of-the-art of nano- and atom photonics are recommended to refer to the literature [2–4, 7–9, 33–37].

Theoretically, we need a reasonable explanation of why light waves or photons are localized, and present an example of an empirical method here. The evanescent field generated on a planar dielectric surface can be obtained by analytical continuation of the wavevector $k_\perp$ normal to the surface as

$\exp(\mathrm{i}\boldsymbol{k}_{\parallel} \cdot \boldsymbol{r} - \gamma z)$ with $k_{\perp} = \mathrm{i}\gamma$ (γ: real), whose localization range is about that of the wavelength of the incident light. Similarly, optical fields around a tiny dielectric sphere that is much smaller than the wavelength are described by the spherical Hankel function with analytic continuation of the radial component of the wavevector $k_{\perp}$ to the pure imaginary $k_{\perp} = \mathrm{i}\gamma$; the first kind of Hankel function of order 0 gives the typical form of an optical near field around a sphere as

$$\phi = \frac{\mathrm{e}^{-\gamma r}}{\gamma r} . \tag{1}$$

The fields satisfying the spherical boundary condition can be expanded in terms of plane waves, or can be expressed using angular spectrum representation. From a numerical analysis, it follows that the distribution of the angular spectrum has a peak at $\gamma = 1/a$ [2]. Applying the result to (1), we see that the localization range of the optical near fields around a sphere is of the order of the radius a, i.e., the size of the material system. Here, it is obvious that an optical near field should be derived consistently from microscopic-induced polarization in matter because they are mutually related, and several approaches in this direction are outlined in the following subsection 1.2.

1.2 Theoretical Approaches

Self-consistent descriptions of optical near fields are required, as mentioned above, and the recent advances in nanoscience and technology have promoted investigation of the dynamics of nanometric material systems interacting with optical near fields, which is a critical and fundamental base for nanophotonics. However, the situation is not simple; an optical near field involves both an optical field and the excitation of matter on a nanometer scale, which is considered as elementary excitation. Moreover, the existence of a boundary or surface strongly affects the elementary excitation mode. It is mandatory to find a proper normal mode, or a suitable basis function to satisfy such boundary conditions for a nanometric system. Such a basis function definitely differs from the plane-wave basis used in bulk theories, but it is very difficult to evaluate it in a rigorous manner except for the simple boundary condition. On a nanometric scale, a quantum nature appears in an optical field–material system, which requires a quantum description of the system creating additional fundamental difficulties.

One possible approach is the semiclassical, self-consistent theory proposed by Cho et al. [38,39], which quantum-mechanically describes microscopic polarization in matter to include a nonlocal response, while the electromagnetic field is classically treated by the Helmholtz equation. They predicted the size dependence and allowance of a dipole-forbidden transition in a nanometric quantum-dot system numerically [40]. Nevertheless, the principal difficulty of quantizing the electromagnetic field using a nonlocal approach remains, as does the difficulty in treating the dynamics of a nanometric material system coupled via optical near fields, because of the complexity involved in a self-consistent procedure.

An alternative approach is to use a normal mode, or the Carniglia–Mandel mode as a complete and orthogonal set that satisfies the infinite planar boundary condition between the dielectric and a vacuum [11,12,23]. This formulation allows us to quantum-field-theoretically describe an evanescent field that is localized along the normal to the surface and possesses an anomalous dispersion relation different from free photons, both of which are important properties of an optical near field. This approach has revealed interesting phenomena that occur near the surface [11, 16–26, 41, 42], but it unfortunately provides neither a self-consistent description of a material system nor a normal mode to satisfy more general or realistic boundary conditions, such as an optical near-field probe–sample system.

Kobayashi et al. [43–45] proposed another approach for focusing on the relevant subsystem that we are interested in, and to extract a characteristic feature of the subsystem after renormalizing the other irrelevant subsystem. Using a quasiparticle representation for a macroscopic material–electromagnetic field system, which only considers the near-field configuration of the system without rigorously taking into account the boundary condition, they showed the localization of optical near fields around a nanometric material system. They have also discussed the dynamics of a single atom, molecule, or nanometric quantum-dot system interacting with the optical near field [46–48].

In this chapter, we present a simple phenomenological model of an optical near field based on this last approach, in order to discuss the dynamic properties of a nanometric material system interacting with an optical near field. The model uses a concept of the "localized photon" to represent the localization property of the optical near field phenomenologically, and two-level systems as representative of a nanometric material system. To clarify the difference between optical near fields and far fields, we compare the results obtained using our model with those of the Dicke model, which consists of a two-level system interacting with radiation field and has been studied extensively since originally presented by Dicke in 1954 [49].

1.3 Difference Between Optical Near Fields and Propagating Fields: Individual vs. Global Excitation

The main difference between a material system coupled with optical near fields and a material system coupled with propagating far fields, is that one is coupled locally and the other is coupled globally, as depicted schematically in Fig. 3.

Since the propagating field is usually expanded in terms of a plane-wave basis, photons emitted as plane waves propagate from one site to distant

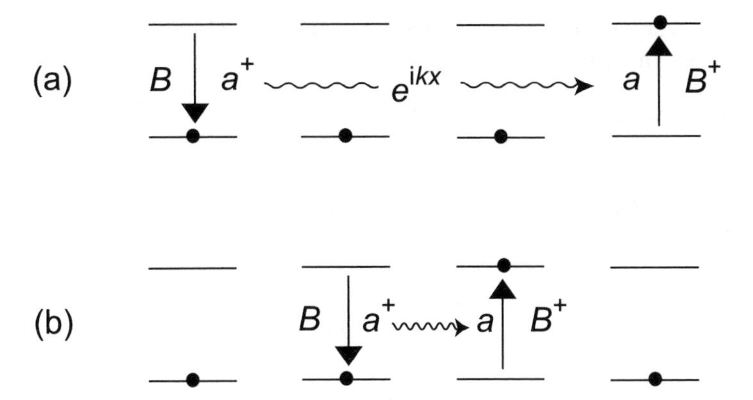

Fig. 3. Schematic drawing of (**a**) a global and (**b**) a local coupling systems. Each element of the system in (**a**) can interact directly with arbitrary elements at a distant site, while the elements at the near site can interact only with each other in (**b**)

sites in the material system and can also excite the far sites (global excitation [50]), as shown in Fig. 3a. By contrast, as shown in Fig. 3b, "near-field photons" or "localized photons" can move only to nearby sites and excite them (individual excitation). Generally speaking, the dynamics of a locally coupled system differ from those of a globally coupled system with respect to the relaxation speed towards equilibrium [51]. Therefore, we expect the dynamics and equilibrium states for a locally coupled system to differ from those of an ordinary globally coupled system, such as a propagating field–matter system [52]. From this perspective, it is important to investigate the nanometric material system that is interacting with optical near fields via individual excitations [46–48, 53].

This chapter presents a model of a material system, i.e., for a quantum-dot system, interacting with optical near fields, and discusses the dynamics of the electric dipole moment and the radiation properties of the system in detail. For comparison, a material system interacting with a radiation field is also examined. We study the propagation of an initial excitation signal in a quantum dot that is prepared by an optical near field locally and individually, and describe it in terms of excitons in N two-level systems. In a dilute limit, excitons are approximated as bosons, and a rigorous solution of the Heisenberg equation shows that the dipole moments representing quantum coherence between any two levels follow linear dynamics. In a nondilute case, excitons at one site obey the fermion commutation relations, while excitons at other sites satisfy the boson commutation relations; this results in nonlinear equations of motion. We predict a coherent oscillation of all the dipoles in the system (dipole ordering), and strong radiation from some of the dipole-ordered states, which are close to Dicke's superradiance [49].

1.4 Two-Level System Interacting with Radiation Fields: Dicke's Superradiance

Dicke originally discussed superradiance, the cooperative emission of radiation from a collection of excited two-level systems, in 1954. He found that under certain conditions the radiation rate is proportional to the square of the number of two-level systems involved, and an intense pulse is emitted [49]. Superradiant phenomena attracted considerable attention from a large number of authors in the 1970s [54–65]. For a small system that is much smaller than the wavelength of the radiation, the equation of motion for the two-level system coupled to the radiation field (Dicke model) is given as follows:

$$\frac{\partial \rho_A}{\partial t} = -\gamma_0 (R_+ R_- \rho_A - 2R_- \rho_A R_+ + \rho_A R_+ R_-) \, , \tag{2}$$

where ρ_A is the density operator after eliminating the radiation field's degrees of freedom from the total density operator of the system, $R_\pm$ is the collective raising and lowering operators of two-level systems [49], and $2\gamma_0$ is the inverse of the spontaneous emission lifetime of each two-level system (Einstein's A coefficient). From (2), one obtains differential equations for the observables of the two-level system. By using a semiclassical approximation of neglecting quantum correlations in the two-level system, an analytical solution [55, 58] for the radiation intensity I can be written as

$$I \propto N^2 \mathrm{sech}^2 \left[\gamma_0 N (t - t_0)\right] \, , \tag{3}$$

which has a cowbell-shaped radiation profile, and shows a radiation pulse emitted under certain initial conditions whose height and width are proportional to N^2 and $1/N$, respectively. Note that the semiclassical approximation does not correctly predict the dynamics starting from completely inverted states [58, 59] with no dipole moments because the quantum fluctuations of the dipole moments are essential in the radiation process.

In a large system, the superradiant state is affected by both inhomogeneity of the two-level system in a sample and the reabsorption of emitted photons [61], which results in decoherence; namely, some part of the coherence in the system is destroyed by the induced dipole–dipole interaction. This reduces peak height and extends the tail of the radiation profile [61,62]. For two-level systems confined within a long cylindrical sample, Bonifacio et al. [60,63] predicted multiple peaks in the radiation profile that stem from the stimulated absorptions and emissions in the long active region. Skribanowitz et al. [64] made experimental observations using a cylindrical sample, and their experimental results were analyzed by Bonifacio et al. [65]. For a small two-level system, the emission spectra of CuCl quantum dots were measured, and Nakamura et al. [66] reported a single peak of pulse emission. Recently, Chen et al. [67] proposed a way to detect a current of superradiance in a double-dot system, while Piovella et al. [68] discussed how to detect a photon echo in the superradiance from a Bose–Einstein condensate.

Tokihiro et al. [69] examined a linear excitonic system, in which excitons can hop from one site to its nearest neighbors via the dipole–dipole interaction, and showed that radiation from a totally inverted state as the initial condition has a reduced peak intensity with extension of the tail, as discussed by Coffey and Friedberg [62]. They also claimed that the radiation profile from a partially excited state shows oscillatory behavior, as indicated by Bonifacio and Lugiato [63]. In addition, T. Brades et al. discussed the oscillatory behavior of a superradiating system coupled to electron reservoirs [70]. It is clear that the peak intensity reduction of radiation in both cases comes from the dipole–dipole decoherence, but it is not obvious whether the multiple pulses so generated have the same origin. Our model also shows multiple pulse emission, and we discuss its mechanism.

1.5 Chapter Outline

The chapter is organized as follows. Section 2 presents a model of N two-level quantum-dot systems interacting with optical near fields represented in terms of localized photons, and describes the model Hamiltonian. Section 3 obtains a rigorous solution of the equations of motion for the system within a boson approximation. Using the solution, we investigate transfer of the dipole moments of the system. In Sect. 4 we obtain the second-order perturbative solution without using the boson approximation to show the dynamics of the system. Section 5 introduces an effective Hamiltonian by renormalizing the degrees of freedom of localized photons. With the help of the effective Hamiltonian, we classify the quasisteady states of the dipole distribution in order to discuss the origin of the dipole ordering. Section 6 is devoted to a review of the Dicke model and superradiance for comparison in the following sections. Section 7 examines the radiation property of the system in a weak limit, where the radiation field does not affect the dynamics of the system. In order to discuss the radiation property of the dissipative system, a quantum master equation is derived in Sect. 8 after introducing a radiation reservoir, and this is solved with and without including quantum correlations between excitons and localized photons. Section 9 deals with the dynamics of localized photons in an open system, which predicts two kinds of phases: storage and nonstorage modes of localized photons in the system. Due to the nonlinearity of the equations of motion, chaotic behavior is seen in the coupling constant dependence of the total number of localized photons. Finally, in Sect. 10 concluding remarks are provided.

2 Model Hamiltonian

One of the most important features of optical near fields is the localization property, where it is not suitable to use a broad spreading wave as a basis function of quantization of the fields. Therefore, it is important to find a good normal mode of the electromagnetic fields to satisfy a boundary condition specific to the problem, by which the optical near fields are produced and quantized [11, 43, 44]. It is difficult, however, to find a general and appropriate normal mode satisfying any arbitrary boundary conditions. As an alternative approach, it is possible to model optical near fields and their important characters phenomenologically. Such an approach is adopted here to formulate the problem and to discuss the dynamics of a nanoscale material system.

To describe the localizability of the optical near fields it is very effective to use a localized basis function such as a Wannier function or a delta function instead of a plane wave. However, at the same time we need to describe the property of a short-range interaction among material sites via optical near fields, or a steep gradient of optical near fields. Taking into account such circumstances we model optical near fields phenomenologically in terms of localized photons that are described as on harmonic oscillator localized in each quantum-dot site, and are only allowed to hop from one site to the nearest neighbors [71]. Figure 4 schematically describes our model system.

We suppose a closely located quantum-dot chain as a nanoscale material system that is expressed as a one-dimensional N two-level system, or an excitonic system with a periodic boundary condition. Since each exciton can only interact with localized photons in the same QD, each quantum dot indirectly interacts with one another via localized photons, as illustrated in Fig. 4. This model is based on the unique property of localization or nonpropagation of optical near fields mentioned above [44], and is preferable for an intuitive understanding of coherent excitation transfer between the

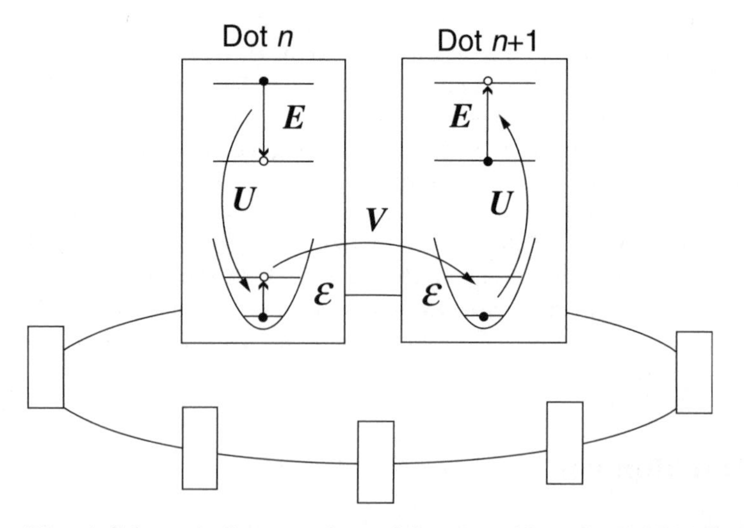

Fig. 4. Schematic diagram of a model system: N two-level QDs labelled by the site number are closely located in a ring, and are interacting with localized photons

QDs and its manipulation by the localized photons. The model Hamiltonian of the system can be written as

$$H = H_{\mathrm{a}} + H_{\mathrm{b}} + H_{\mathrm{int}} , \tag{4}$$

where H_{a} represents localized photons, H_{b} describes excitons, and H_{int} represents the localized photon–exciton interaction. Each Hamiltonian can be expressed as

$$H_{\mathrm{a}} = \sum_{n=1}^{N} \left\{ \varepsilon a_n^{\dagger} a_n + V (a_{n+1}^{\dagger} a_n + a_n^{\dagger} a_{n+1}) \right\} , \tag{5a}$$

$$H_{\mathrm{b}} = E \sum_{n=1}^{N} b_n^{\dagger} b_n , \tag{5b}$$

$$H_{\mathrm{int}} = U \sum_{n=1}^{N} (a_n^{\dagger} b_n + b_n^{\dagger} a_n) , \tag{5c}$$

where n indicates the site number, and a_n $(a_n^{\dagger})$ and b_n $(b_n^{\dagger})$ represent annihilation (creation) operators of a localized photon and an exciton, respectively. The periodic boundary condition requires that the $(N+1)$-th site corresponds to the first site. The constant energies of the localized photons and excitons are represented by $\varepsilon = \hbar\omega$ and $E = \hbar\Omega$, respectively. The hopping energy of the localized photons is represented as $V = \hbar v$, and $U = \hbar g$ gives the strength of the conventional dipolar coupling between a localized photon and an exciton in the rotating-wave approximation.

We apply the boson commutation relations to the localized photons as

$$[a_n, a_{n'}^{\dagger}] = \delta_{nn'} , \quad [a_n, a_{n'}] = [a_n^{\dagger}, a_{n'}^{\dagger}] = 0 . \tag{6}$$

The creation operator of an exciton, $b_n^{\dagger}$, can be written in terms of the annihilation operator of a valence electron of the n-th site, c_{n0}, and the creation operator of a conducting electron of n-th site, $c_{nf}^{\dagger}$, as $b_n^{\dagger} = c_{nf}^{\dagger} c_{n0}$. In the same way, the annihilation operator of an exciton, b_n, can be written as $b_n = c_{n0}^{\dagger} c_{nf}$. Assuming that only one exciton is generated in a quantum dot site, we obtain the relation as $N_f + N_0 = c_{nf}^{\dagger} c_{nf} + c_{n0}^{\dagger} c_{n0} = 1$. With the help of these expressions and the fermi commutation relations for electrons, the commutation relations for excitons [72, 73] are derived as

$$[b_{n'}, b_n^{\dagger}] = \delta_{nn'} (1 - 2 b_n^{\dagger} b_n) , \tag{7}$$

which shows that excitons behave as fermions at intrasite locations and as bosons at intersite ones. It follows from (7) that excitons are approximated as bosons in a dilute limit of the exciton density, $\langle b_n^{\dagger} b_n \rangle \equiv \langle N_n \rangle \ll 1$.

With the boson approximation, the Heisenberg equation of motion can be rigorously solved. First, we obtain a rigorous solution of the Heisenberg equation for bosonic excitons to investigate the dynamical properties. Then we investigate the dynamics of fermionic excitons, solving the Heisenberg equation both perturbatively and numerically.

3 Dynamics of Bosonic Excitons

3.1 Boson Approximation and Diagonalization of the Hamiltonian

To solve the Heisenberg equation we introduce the spatial Fourier transformation for a_n and b_n as

$$A_k = \frac{1}{\sqrt{N}} \sum_{n=1}^{N} \mathrm{e}^{ikn} a_n \, , \tag{8a}$$

$$B_k = \frac{1}{\sqrt{N}} \sum_{n=1}^{N} \mathrm{e}^{ikn} b_n \, , \tag{8b}$$

where we set $k = 2\pi l/N$ for $l = 1, \cdots, N$ and the site distance as 1. Fourier transforms A_k satisfy the following commutation relations

$$[A_k, A_{k'}^\dagger] = \delta_{kk'} \, , \quad [A_k, A_{k'}] = [A_k^\dagger, A_{k'}^\dagger] = 0 \, , \tag{9}$$

and the commutation relations of Fourier transforms B_k are similarly obtained as

$$[B_k, B_{k'}^\dagger] = \frac{1}{N} \sum_{n,n'} \mathrm{e}^{\mathrm{i}(kn-k'n')} [b_n, b_{n'}^\dagger] = \frac{1}{N} \sum_{n} \mathrm{e}^{\mathrm{i}(k-k')n} (1 - 2b_n^\dagger b_n) \, . \tag{10}$$

It follows from this expression that when the expectation value of the number of excitons is small as $\langle b_n^\dagger b_n \rangle \ll N$, the second term on the right-hand side of (10) can be neglected, and excitons are approximated as bosons as

$$[B_k, B_{k'}^\dagger] = \delta_{kk'} \, , \quad [B_k, B_{k'}] = [B_k^\dagger, B_{k'}^\dagger] = 0 \, , \tag{11}$$

where the following relation

$$\frac{1}{N} \sum_{n} \mathrm{e}^{\mathrm{i}(k-k')n} = \delta_{kk'} \tag{12}$$

is used. Fourier inverse transforms are also given as

$$a_n = \frac{1}{\sqrt{N}} \sum_{k \in 1BZ} \mathrm{e}^{-ikn} A_k \, , \tag{13a}$$

$$b_n = \frac{1}{\sqrt{N}} \sum_{k \in 1BZ} \mathrm{e}^{-ikn} B_k \, , \tag{13b}$$

where the summation of k runs over the first Brillouin zone ($1BZ$). Using (8a) and (8b), and the following relation

$$\frac{1}{N} \sum_{k \in 1BZ} \mathrm{e}^{\mathrm{i}k(n-n')} = \delta_{nn'} \, , \tag{14}$$

we can transform the Hamiltonian (4) into a Fourier transform representation

$$H = \sum_k \{(\varepsilon + 2V \cos k) A_k^\dagger A_k + E B_k^\dagger B_k + U(A_k B_k^\dagger + A_k^\dagger B_k)\}$$
$$\equiv \sum_k H_k \,. \tag{15}$$

Note that the dispersion relation of localized photons is changed from ε to $(\varepsilon + 2V \cos k)$. The transformed Hamiltonian H_k can be diagonalized into a simple quadratic form

$$H_k = (A_k^\dagger, B_k^\dagger) \begin{pmatrix} \varepsilon + 2V \cos k & U \\ U & E \end{pmatrix} \begin{pmatrix} A_k \\ B_k \end{pmatrix}$$
$$\equiv (A_k^\dagger, B_k^\dagger) \begin{pmatrix} e & U \\ U & E \end{pmatrix} \begin{pmatrix} A_k \\ B_k \end{pmatrix} \equiv \sum_{i,j=1}^{2} M_{ij} X^\dagger(k)_i X(k)_j \,, \tag{16}$$

where the abbreviations $e = \varepsilon + 2V \cos k$ and $\boldsymbol{X}^\dagger(k) = (A_k^\dagger, B_k^\dagger)$ are used. The matrix kernel of the Hamiltonian, M, is diagonalized as

$$S^T M S = \begin{pmatrix} \lambda_+ & 0 \\ 0 & \lambda_- \end{pmatrix} \,, \tag{17}$$

by an orthogonal matrix S as

$$S = \begin{pmatrix} \sqrt{\frac{L+K}{2L}} & \sqrt{\frac{L-K}{2L}} \\ \sqrt{\frac{L-K}{2L}} & -\sqrt{\frac{L+K}{2L}} \end{pmatrix} \,. \tag{18}$$

Here, eigenvalues $\lambda_\pm$ are written as

$$\lambda_\pm = \frac{e+E}{2} \pm \frac{1}{2}\sqrt{(e-E)^2 + 4U^2}$$
$$\equiv K + E \pm L \,, \tag{19}$$

with $K = (e - E)/2$ and $L = \sqrt{K^2 + U^2}$. Using the polariton transformation as

$$\begin{pmatrix} \alpha_k \\ \beta_k \end{pmatrix} \equiv S^T \begin{pmatrix} A_k \\ B_k \end{pmatrix} \,, \tag{20}$$

we can finally diagonalize the Hamiltonian (15) as follows:

$$H = \sum_k H_k = \sum_k (\lambda_+ \alpha_k^\dagger \alpha_k + \lambda_- \beta_k^\dagger \beta_k) \,. \tag{21}$$

3.2 Dipole Dynamics Driven by Local Excitation

With the help of the diagonalized Hamiltonian (21), we can immediately solve the Heisenberg equation and can express the time evolution of exciton-polariton operators $\alpha_k(t)$ and $\beta_k(t)$ as

$$\begin{pmatrix} \alpha_k(t) \\ \beta_k(t) \end{pmatrix} = \begin{pmatrix} e^{-i\lambda_+ t/\hbar}\alpha_k \\ e^{-i\lambda_- t/\hbar}\beta_k \end{pmatrix} , \tag{22}$$

where α_k and β_k denote the operators in the Schrödinger representation. It is assumed in this chapter that an operator with no indication of the time dependence is expressed in the Schrödinger picture as a time-independent operator. From the inverse transformation of (20), we obtain a time-dependent solution of the Fourier transforms $A_k(t)$ and $B_k(t)$ as

$$\begin{pmatrix} A_k(t) \\ B_k(t) \end{pmatrix} = S \begin{pmatrix} \alpha_k(t) \\ \beta_k(t) \end{pmatrix} = S \begin{pmatrix} e^{-i\lambda_+ t/\hbar}\alpha_k \\ e^{-i\lambda_- t/\hbar}\beta_k \end{pmatrix} . \tag{23}$$

By transforming the exciton-polariton operators (α_k, β_k) into (A_k, B_k) again, explicit time-dependent solutions $A_k(t)$ and $B_k(t)$ of the Heisenberg equations are written as

$$\begin{aligned} A_k(t) &= e^{-i(e+E)t/2\hbar}\left(\cos\frac{L}{\hbar}t - i\frac{K}{L}\sin\frac{L}{\hbar}t\right)A_k \\ &\quad - i\frac{U}{L}e^{-i(e+E)t/2\hbar}\left(\sin\frac{L}{\hbar}t\right)B_k , \end{aligned} \tag{24a}$$

$$\begin{aligned} B_k(t) &= -i\frac{U}{L}e^{-i(e+E)t/2\hbar}\left(\sin\frac{L}{\hbar}t\right)A_k \\ &\quad + e^{-i(e+E)t/2\hbar}\left(\cos\frac{L}{\hbar}t + i\frac{K}{L}\sin\frac{L}{\hbar}t\right)B_k . \end{aligned} \tag{24b}$$

Using the Fourier inverse transformation, we obtain a final expression of the time evolution of the exciton operator $b_n(t)$ as follows:

$$\begin{aligned} b_n(t) &= \frac{1}{N}\sum_{k,m} e^{ik(m-n)-i(e+E)t/2\hbar} \\ &\quad \times \left\{-ia_m\frac{U}{L}\sin\frac{L}{\hbar}t + b_m\left(\cos\frac{L}{\hbar}t + i\frac{K}{L}\sin\frac{L}{\hbar}t\right)\right\} . \end{aligned} \tag{25}$$

Here we introduce a physical quantity of the electric dipole moment for our two-level system defined as

$$\boldsymbol{\mu}_n(t) = \boldsymbol{\mu}(b_n(t) + b_n^\dagger(t)) = \boldsymbol{\mu}P_n(t) , \tag{26}$$

where $P_n(t) = b_n(t) + b_n^\dagger(t)$ denotes the dipole operator that describes the electric dipole moment of each quantum dot. Moreover, two other independent operators are defined as

$$V_n(t) = i(b_n(t) - b_n^\dagger(t)) , \qquad W_n(t) = b_n^\dagger(t)b_n(t) - b_n(t)b_n^\dagger(t) , \tag{27}$$

where the latter represents the population difference of the system. The unit operator and these three operators form a basis for any two-level system, and the set (P_n, V_n, W_n) corresponds to the Bloch vector in a collection of two-level atom systems. However, it should be noted that the population difference W_n is always -1 for bosonic excitons, and the number operators of excitons $b_n^\dagger b_n$ are the independent operators in this case.

In order to investigate the coherent excitation dynamics of the system, we examine the time evolution of the expectation value of the dipole moment at an arbitrary QD n, $\langle P_n(t)\rangle$, under a variety of initial conditions. The expression given by (25) and its Hermitian conjugate provide a useful result as

$$\begin{aligned}\langle P_n(t)\rangle = \frac{1}{N}\sum_{k\in 1BZ}\sum_{m=1}^{N}&\left\{\left(\langle P_m\rangle\cos\xi_{mn} + \langle V_m\rangle\sin\xi_{mn}\right)\cos\frac{L}{\hbar}t\right.\\ &\left.+\frac{K}{L}\left(\langle V_m\rangle\cos\xi_{mn} - \langle P_m\rangle\sin\xi_{mn}\right)\sin\frac{L}{\hbar}t\right\}\,, \end{aligned}\tag{28}$$

where the operation $\langle\cdots\rangle = \mathrm{Tr}\rho\cdots$ means the expectation value of an arbitray operator. The notation $\xi_{mn} \equiv k(m-n) - t(e+E)/2\hbar$ is used. Localized photons are assumed to be initially in the vacuum state. In particular, it is intriguing to investigate the dynamics with the initial condition of a locally or individually excited state, i.e., how the initial excitation prepared only at one site is transferred in the system. Setting $\langle P_n(0)\rangle = \delta_{n1}\langle P_1\rangle$ and $\langle V_n(0)\rangle = 0$ for all n, we obtain from (28) an explicit solution as

$$\langle P_n(t)\rangle = \frac{1}{N}\sum_{k}\left\{\cos\xi_{1n}\cos\frac{L}{\hbar}t - \frac{K}{L}\sin\xi_{1n}\sin\frac{L}{\hbar}t\right\}\langle P_1\rangle\,. \tag{29}$$

In Fig. 5 we present one of the numerical results of the time evolution of the dipole moment distribution calculated from (29), where the total number of sites is eight, and the parameter values $E = 2$, $\varepsilon = 1$, and $V = 1$ are used in a unit of a typical exciton energy in the QD.

Although the dynamics seems complicated at first sight, we can find a characteristic behavior that the dipole of the first site moves to the fifth site (the opposite site) and returns to the first site again. To see the behavior more precisely we show plots of the time evolution of the dipole moments at the first site and the fifth site in Fig. 6a and Fig. 6b, respectively.

The result clearly shows as a dominant behavior that when the dipole of the first site is active, that of the fifth site is inactive and vice versa, or the dipole moves like a seesaw between the first site and the fifth site. The recurrence behavior is reasonable from the fact that the Heisenberg equation of the system is linear in the boson approximation. Since the dipole moment P_n is quantum coherence between the ground and excited states of a two-level system, quantum coherence is transported to the farthest site of the chain and then comes back to the original site. When the total number of sites is odd, the seesaw motion arises between an initially excited site and its farthest pair of sites of the chain. An interesting point in this case is that we

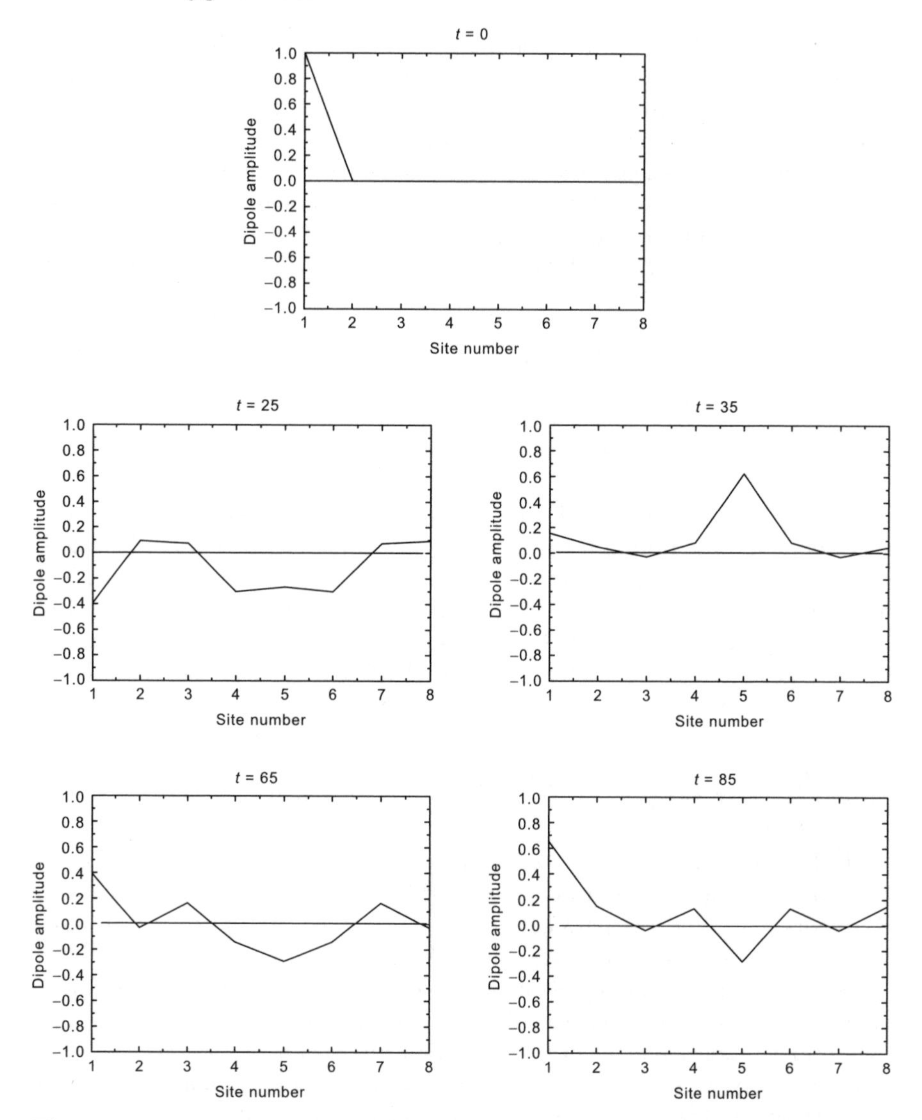

Fig. 5. Time evolution of the dipole-moment distribution when the total number of sites is eight. The parameters $E = 2$, $\varepsilon = 1$, and $V = 1$ are used in the calculation. The vertical and horizontal axes represent the dipole amplitude and the site number, respectively. The dipole moment is initially set only at the first site

can obtain two copies of coherence at the farthest pair of sites as illustrated in Fig. 7b.

Moreover, it might be possible to use the system as a nanophotonic device of transporting or splitting quantum coherence.

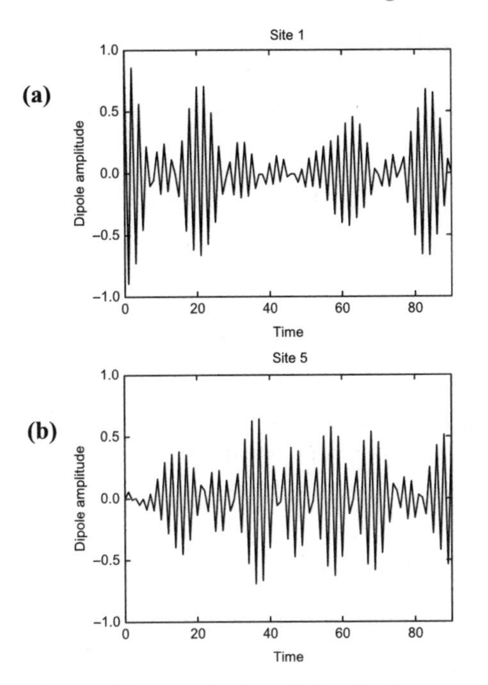

Fig. 6. Time evolution of the dipole moment of (**a**) the first site and (**b**) the fifth site. When the dipoles are large in (**a**), corresponding dipoles in (**b**) are small, and vice versa

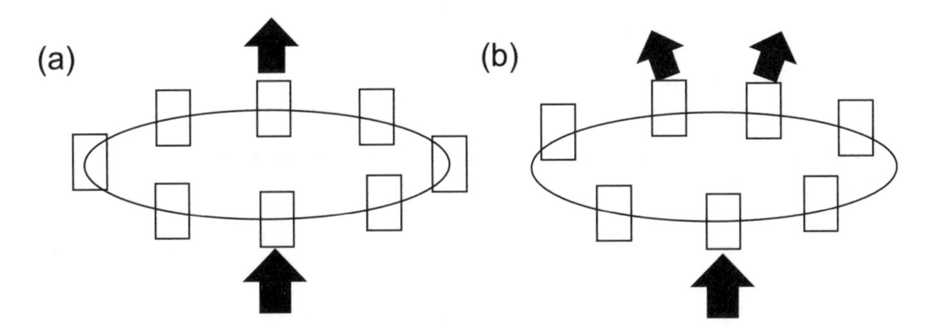

Fig. 7. Transportation of quantum coherence in (**a**) an even site number system and (**b**) an odd-site number system. In the even-site system (**a**), output signals are obtained from the opposite site to the input site. In the odd-site system (**b**), output signals are split into the two-pair site opposite to the input site

4 Dynamics of Fermionic Excitons

If we rigorously adopt the commutation relation for excitons given by (7), the Heisenberg equations for the system read

$$\frac{\mathrm{d}b_n(t)}{\mathrm{d}t} = \frac{\mathrm{i}}{\hbar}[H, b_n(t)] = -\frac{\mathrm{i}}{\hbar}Eb_n(t) + \frac{\mathrm{i}}{\hbar}Ua_n(t)W_n(t) \ , \tag{30a}$$

$$\begin{aligned}\frac{\mathrm{d}a_n(t)}{\mathrm{d}t} &= \frac{\mathrm{i}}{\hbar}[H, a_n(t)] \\ &= -\frac{\mathrm{i}}{\hbar}\varepsilon a_n(t) - \frac{\mathrm{i}}{\hbar}V(a_{n+1}(t) + a_{n-1}(t)) - \frac{\mathrm{i}}{\hbar}Ub_n(t) \ . \end{aligned} \tag{30b}$$

Since the higher-order terms are produced by a mode–mode coupling such as $a_n W_n = a_n(b_n^\dagger b_n - b_n b_n^\dagger)$, the equations become nonlinear and are hardly solved analytically. Thus we first solve the equations perturbatively in order to investigate the dynamics discussed in the preceding section.

4.1 Perturbative Expansion of Time-Evolution Operator

Noticing that the Hamiltonian for localized photons H_a given by (5a) can be written in a quadratic form as

$$H_a = (a_1^\dagger, \cdots, a_N^\dagger) \begin{pmatrix} \varepsilon & V & & & V \\ V & \varepsilon & V & & \\ & \cdots\cdots & & & \\ & & \cdots\cdots & & \\ & & V & \varepsilon & V \\ V & & & V & \varepsilon \end{pmatrix} \begin{pmatrix} a_1 \\ \vdots \\ \vdots \\ a_N \end{pmatrix} \tag{31}$$

$$\equiv \sum_{nm} a_n^\dagger R_{nm} a_m \ , \tag{32}$$

we obtain an orthogonal matrix P to diagonarize the matrix kernel R of the Hamiltonian as

$$(P^{-1}RP)_{ij} = \hbar\lambda_i\delta_{ij} \ . \tag{33}$$

Then new annihilation operators $v_j = \sum_n P_{nj}a_n$ and creation operator $v_j^\dagger = \sum_n P_{nj}a_n^\dagger$ of new modes of localized photons are introduced. Here note that the new modes are extended over a whole system, since all the elements of the diagonalization matrix P are not zero. The commutation relation of the new operators can be given as $[v_i, v_j^\dagger] = \delta_{ij}$. With the help of (33) and the modes v_i, the Hamiltonian in (32), or (5a) is diagonalized as

$$H_\mathrm{a} = \sum_j \hbar\lambda_j v_j^\dagger v_j \ , \tag{34}$$

and the localized photon–exciton interaction $H_{\rm int}$ in (5c) is written in terms of v_i as

$$H_{\rm int} = \hbar g \sum_{nj} P_{nj} (v_j^\dagger b_n + v_j b_n^\dagger) \, . \tag{35}$$

In order to derive a perturbative expansion of the time evolution of an arbitrary operator, we define the Liouvillians as

$$L \cdots = \frac{1}{\hbar}[H, \cdots] \, , \quad L_0 \cdots = \frac{1}{\hbar}[H_0, \cdots] \, , \quad L_{\rm int} \cdots = \frac{1}{\hbar}[H_{\rm int}, \cdots] \, . \tag{36}$$

Then the Heisenberg equation of an arbitrary operator O is written as

$$\dot{O} = \frac{\rm i}{\hbar}[H, O] = {\rm i}LO \, , \tag{37}$$

and we can obtain a formal solution as

$$O(t) = G(t)O(0) \, , \tag{38}$$

with the time evolution operator $G(t) = {\rm e}^{{\rm i}Lt}$ that satisfies the following equation

$$\dot{G}(t) = {\rm i}LG(t) = {\rm i}L_0 G(t) + {\rm i}L_{\rm int} G(t) \, . \tag{39}$$

Treating the interaction term $H_{\rm int}$ as a perturbation, we solve (39) perturbatively up to the second order of the perturbation as [73]

$$\begin{aligned} G^{(2)}(t) = G_0(t) &+ {\rm i} \int_0^t G_0(t-s) L_{\rm int} G_0(s) {\rm d}s \\ &- \int_0^t {\rm d}s \int_0^s {\rm d}u G_0(t-s) L_{\rm int} G_0(s-u) L_{\rm int} G_0(u) \, , \end{aligned} \tag{40}$$

where the notation $G_0(t) = {\rm e}^{{\rm i}L_0 t}$ is used. Substituting (40) into (38) and using (36), we can obtain the time evolution of the exciton operator $b_n(t) = G^{(2)}(t) b_n$. Suppose that localized photons are initially in the vacuum and $\langle V_n \rangle = 0$, then the expectation value of $b_n(t)$ is expressed as

$$\begin{aligned} \langle b_n(t) \rangle = {\rm e}^{-{\rm i}\Omega t} \langle b_n \rangle &+ g^2 \sum_{j=1}^N \Gamma_j(t) \\ &\times \left(\sum_{m=1}^N P_{nj} P_{mj} (1 - \delta_{nm}) \langle W_n \rangle \langle b_m \rangle - P_{nj}^2 \langle b_n \rangle \right) \, , \end{aligned} \tag{41}$$

where the notation

$$\begin{aligned} \Gamma_j(t) &= {\rm e}^{-{\rm i}\Omega t} \int_0^t {\rm d}s {\rm e}^{{\rm i}s(\Omega - \lambda_j)} \int_0^s {\rm d}u {\rm e}^{-{\rm i}u(\Omega - \lambda_j)} \\ &= \frac{{\rm e}^{-{\rm i}\Omega t} - {\rm e}^{-{\rm i}\lambda_j t}}{(\Omega - \lambda_j)^2} + \frac{{\rm i}t{\rm e}^{-{\rm i}\Omega t}}{\Omega - \lambda_j} \equiv c_j(t) + {\rm i}d_j(t) \end{aligned} \tag{42}$$

is used, and $c_j(t)$ and $d_j(t)$ represent the real and imaginary parts of $\Gamma_j(t)$, respectively. The expression given by (41) and its Hermitian conjugate provide a time evolution of the expectation value of the dipole at an arbitrary QD site n as

$$\langle P_n(t)\rangle = \langle P_n\rangle \left(\cos\Omega t - g^2 \sum_j c_j(t) P_{nj}^2\right) + g^2 \sum_j \sum_{m\neq n} c_j(t) P_{nj} P_{mj} \langle P_m\rangle \langle W_n\rangle \, . \tag{43}$$

Corresponding to the discussion developed in Sect. 3.2, we set initially $\langle P_n\rangle = \delta_{n1}$ and obtain the perturbative solution that describes the dipole dynamics driven by a local excitation of QD site one as follows:

$$\langle P_n(t)\rangle = \delta_{n1} \left(\cos\Omega t - g^2 \sum_j c_j(t) P_{nj}^2\right) \langle P_1\rangle + (1-\delta_{n1}) g^2 \sum_j c_j(t) P_{nj} P_{1j} \langle P_1\rangle \langle W_n\rangle \, . \tag{44}$$

In the next section we discuss the dynamical properties of the system, on the basis of the solution given by (44).

4.2 Numerical Results and Dynamical Properties

As discussed in Sect. 3.2, we numerically examine the dynamics of the dipole of the system depending on the initial conditions. Suppose that localized photons are initially in the vacuum, $\langle V_n\rangle = 0$ for all n, the dipole is initially set only at the first site as $\langle P_n\rangle = \delta_{n1}$, and all the populations are in the ground states except for the first site as $\langle W_n\rangle = -(1-\delta_{n1})$. As mentioned before, all the population differences are automatically -1 for bosonic excitons. In addition, we impose that the length of the Bloch vector (P_n, V_n, W_n) is normalized as

$$\langle P_n(t)\rangle^2 + \langle V_n(t)\rangle^2 + \langle W_n(t)\rangle^2 = 1 \, . \tag{45}$$

In Fig. 8 the time evolution of the dipole distribution is plotted. The parameter values used are the same as in Sect. 3.2 ($E = 2$, $\varepsilon = 1$, and $V = 1$).

It follows from the figure that the dipole excitation of the first site moves to the opposite site as shown in the boson case (see Fig. 5). The reason is that the initial exciton density is so dilute as to validate the boson approximation. As time advances, the amplitude of each dipole increases because the perturbative solution violates the unitarity.

To find a new feature of fermionic excitons, we vary the initial conditions for the population differences from $\langle W_n\rangle = -(1-\delta_{n1})$. Since the initial

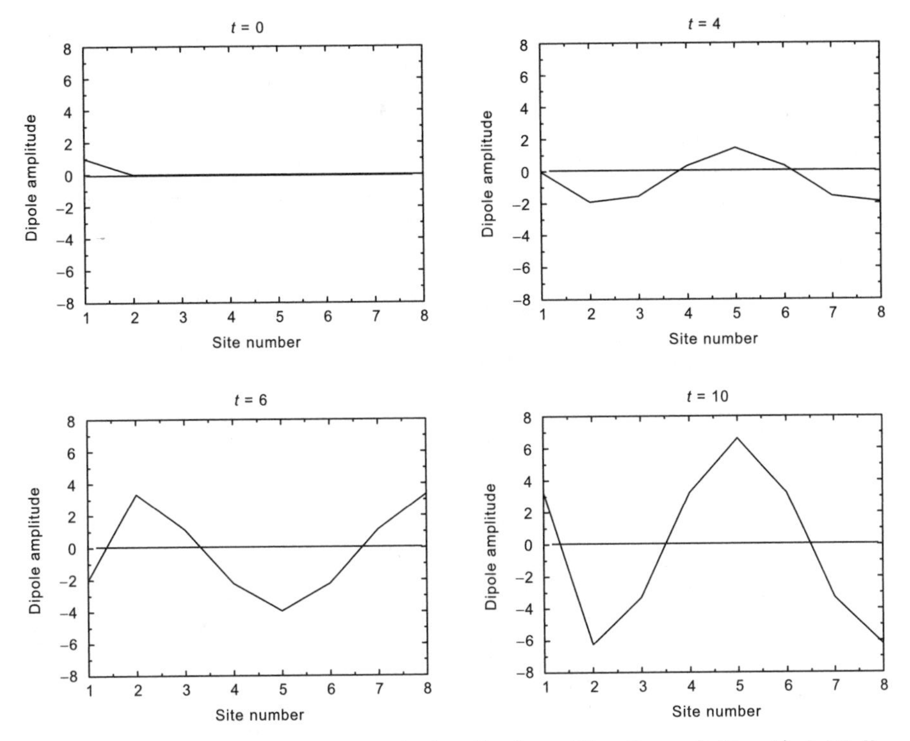

Fig. 8. Time evolution of the dipole distributions ($E = 2, \varepsilon = 1, V = 1$). Initially all of the sites except the first site are in the ground state, while the dipole moment is initially set only at the first site

population differences at site n contributes to the perturbative solution of the dipole at the same site, (44), in a product of $\langle W_n \rangle \langle P_1 \rangle$, the following hypothesis is proposed.

Flip Hypothesis

If the sign of the initial population difference of the n-th site, $\langle W_n(0) \rangle$, is inverted, then the direction of the dipole moment of the n-th site at arbitrary time, $\langle P_n(t) \rangle$, is flipped.

Since the dipoles at sites 4, 5 and 6 in Fig. 8 are directed opposite to the others, we invert the sign of $\langle W_n \rangle$ for $n = 4, 5, 6$. Figure 9 shows the result that the direction of the dipoles at sites 4, 5, and 6 are flipped, and that all the dipoles oscillate with the same phase but with different amplitudes. Thus we observe that the system is transferred from a locally excited state to a coherently oscillating state of the dipoles, in other words, to a dipole-ordered state. In the ordered state the total dipole is N times larger than each single dipole. Moreover, since radiation from an oscillating dipole is proportional to

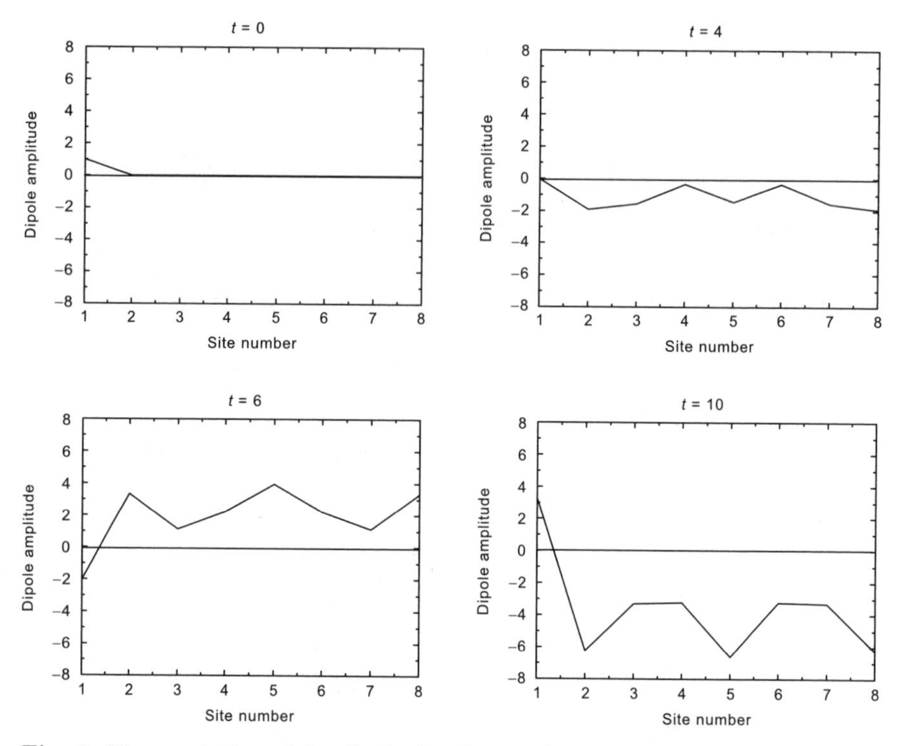

Fig. 9. Time evolution of the dipole distribution ($E = 2, \varepsilon = 1, V = 1$): the initial population differences of sites 4, 5 and 6 are inverted from $\langle W_n \rangle = -(1 - \delta_{n1})$

the square of the dipole moment, we can expect a high intensity of radiation from the dipole-ordered state. The radiation property of the system will be discussed in Sect. 7.1.

4.3 Dynamics of Dipole-Forbidden States via Optical Near-Field Interaction

It is well known that the electric interaction between molecules with no dipole moments is weak when they are separated in a macroscopic distance, and that the interaction becomes strong when they are very close to each other. Taking this into account, we investigate the dynamics of the system driven by localized photons, from an initial state that alternating dipoles are set and thus the total dipole of the system vanishes for an even number of sites (see Fig. 10a). Such a dipole-forbidden state can be manipulated by localized photons, not by propagating far fields.

As an initial condition we set $\langle P_n \rangle = (-1)^n$, and for simplicity, localized photons are in the vacuum, $\langle V_n \rangle = 0$, and $\langle W_n \rangle = 0$. The system parameters $E = 2$, $\varepsilon = 1$, and $V = 1$ are used as before. Figure 11 shows the time evolution of such a dipole-forbidden state.

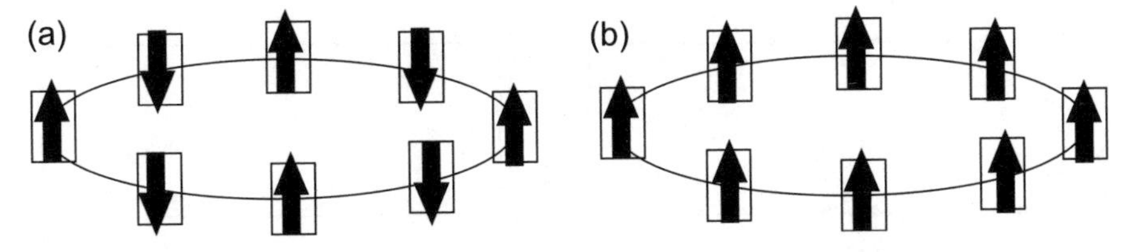

Fig. 10. Schematic illustration of (**a**) an alternating dipole distribution (a dipole-forbidden state) and (**b**) a dipole-ordered state. The state with alternating dipoles that result in the total dipole of zero cannot be coupled by the radiation field with the dipole interaction

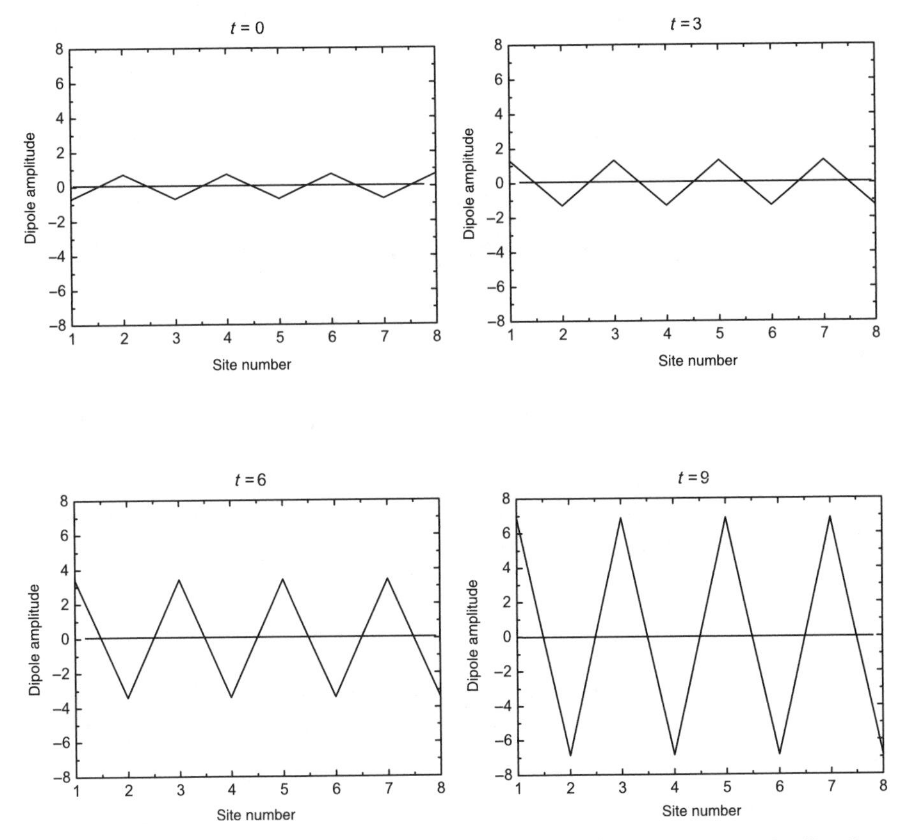

Fig. 11. Time evolution of a dipole-forbidden state (alternating dipole distribution). All the population differences are initially set as 0. We observe an oscillation of the alternating dipoles. The system remains in the dipole-forbidden state

From Fig. 11 it follows that the system oscillates as schematically shown in Fig. 10a, and that it remains in the initial dipole-forbidden state. This kind of dynamics is achieved in the system whose initial distribution of the population differences is uniform. For example, if we set $\langle P_n \rangle = (-1)^n/\sqrt{2}$

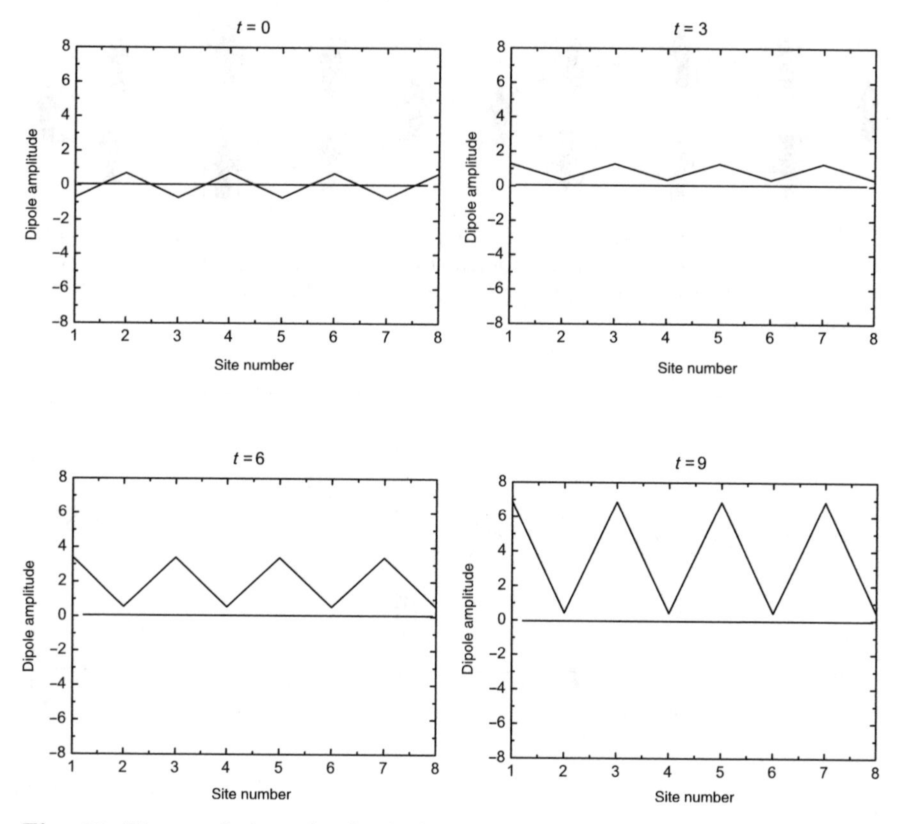

Fig. 12. Time evolution of a dipole-forbidden state ($E = 2, \varepsilon = 1, V = 1$). The population differences are initially set as $\langle W_n \rangle = -\langle P_n \rangle$. The system is converted from a dipole-forbidden state to a dipole-ordered state

and $\langle W_n \rangle = 1/\sqrt{2}$ for all n, the result is the same as shown in Fig. 11 except for its amplitude of the oscillation.

Next, we manipulate the distribution of the population difference nonuniformly, so that the signs are set as opposite to those of the corresponding dipoles:

$$\langle P_n \rangle = (-,+,-,+,-,+,-,+)/\sqrt{2}\,,$$
$$\langle W_n \rangle = (+,-,+,-,+,-,+,-)/\sqrt{2}\,.$$

Figure 12 presents the result that a dipole-forbidden state as shown in Fig. 10a is converted to a dipole-ordered state as illustrated in Fig. 10b.

Since the dipole ordering has occurred by manipulating the initial distribution of the population differences, the result can be interpreted by the "*flip hypothesis*" proposed in Sect. 4.2. As a collective oscillation of the dipoles occurred, the system evolved from a nonradiative state to a radiative state through the localized photon–exciton interaction.

4.4 Semiclassical Approximation

Here we should note that the second-order perturbative solutions break the unitarity, and that the long-time behaviors of the dynamics given by the solutions are not necessarily the same as the exact ones. Thus we have to return to the Heisenberg equations, (30a) and (30b), to evaluate the time evolution of physical observables. The Heisenberg equations are first solved by neglecting the quantum correlations between excitons and localized photons such as $\langle W_n y_n \rangle = \langle W_n \rangle \langle y_n \rangle$, and later the quantum correlations are estimated in Sect. 8.2. In this semiclassical approximation, (30a) and (30b) can be converted to the following coupled differential equations

$$\begin{cases} \partial_t \langle P_n \rangle = -\Omega \langle V_n \rangle + g \langle W_n \rangle \langle y_n \rangle \,, \\ \partial_t \langle V_n \rangle = \Omega \langle P_n \rangle - g \langle W_n \rangle \langle x_n \rangle \,, \\ \partial_t \langle W_n \rangle = g \left(\langle V_n \rangle \langle x_n \rangle - \langle P_n \rangle \langle y_n \rangle \right) \,, \end{cases} \tag{46}$$

$$\begin{cases} \partial_t \langle x_n \rangle = -\omega \langle y_n \rangle - v \left(\langle y_{n-1} \rangle + \langle y_{n+1} \rangle \right) - g \langle V_n \rangle \,, \\ \partial_t \langle y_n \rangle = \omega \langle x_n \rangle + v \left(\langle x_{n-1} \rangle + \langle x_{n+1} \rangle \right) + g \langle P_n \rangle \,, \end{cases} \tag{47}$$

where the notations $x_n = a_n + a_n^\dagger$, $y_n = \mathrm{i}(a_n - a_n^\dagger)$, and $v = V/\hbar$ are used.

To check the reliability of the semiclassical approximation, we again investigate the system examined in Sect. 4.2. Figure 13 shows the time evolution of the dipole distribution obtained from the semiclassical approach for the system corresponding to Fig. 8.

Comparing Fig. 13 with Fig. 8, we find that both profiles are the same, and that the amplitude of each dipole obtained from the Heisenberg equations is less than 1, which means that the unitarity of the time evolution is conserved.

In Fig. 14, we show the time evolution of the dipole distribution obtained from the semiclassical approach when the initial population differences of sites 4, 5, and 6 are inverted. It follows from the figure that each dipole is ordered as the flip hypothesis predicts, and that the flip hypothesis is still valid in the numerical solution of the semiclassical Heisenberg equation.

Numerical results obtained from (46) and (47) with a semiclassical approximation predict the dipole dynamics with qualitative similarities as shown from the perturbative solution (44). The flip hypothesis proposed in Sect. 4.2 is also verified by using the numerical solutions of (46) and (47), or those of the Heisenberg equations without quantum correlations. These results show that the approach discussed in this section, maintaining an advantage to conserve the unitarity, qualitatively describe the dipole dynamics of the system in a similar way as the other two approaches. The validity of the approximation employed here, which we will examine in Sect. 8.2, is reported in the Dicke model that superradiant phenomena can be described with a semiclassical approximation that neglects quantum correlations among the atoms [57], and that the contribution from the quantum correlations is of the order of $1/N$ for large N [58].

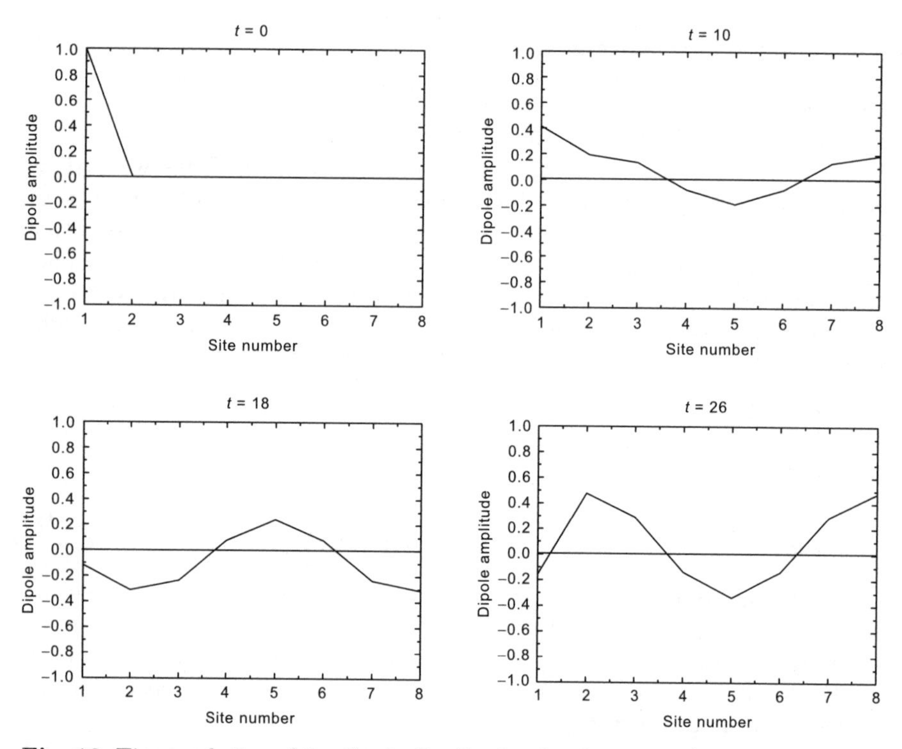

Fig. 13. Time evolution of the dipole distribution for the system ($E = 2, \varepsilon = 1, V = 1$) obtained from the Heisenberg equation with the semiclassical approximation. Initially all of the sites except the first site are in the ground state

5 Effective Hamiltonian and the Dipole Ordering

In order to investigate the origin of the dipole ordering discussed above, we first introduce an effective Hamiltonian to renormalize the degrees of freedom of localized photons. Using the effective Hamiltonian, we then classify quasisteady states with respect to the dipole distribution.

5.1 Effective Hamiltonian

A unitary transformation with an anti-Hermitian operator S is applied to the Hamiltonian (4) as [74–76]

$$\tilde{H} = \mathrm{e}^{-S} H \mathrm{e}^{S} = H + [H, S] + \frac{1}{2}[[H, S], S] + \cdots . \tag{48}$$

If the Hamiltonian (4) is divided into the unperturbed part $H_0 = H_\mathrm{a} + H_\mathrm{b}$ and perturbed part H_int, and if S is chosen to satisfy

$$H_\mathrm{int} + [H_0, S] = 0 , \tag{49}$$

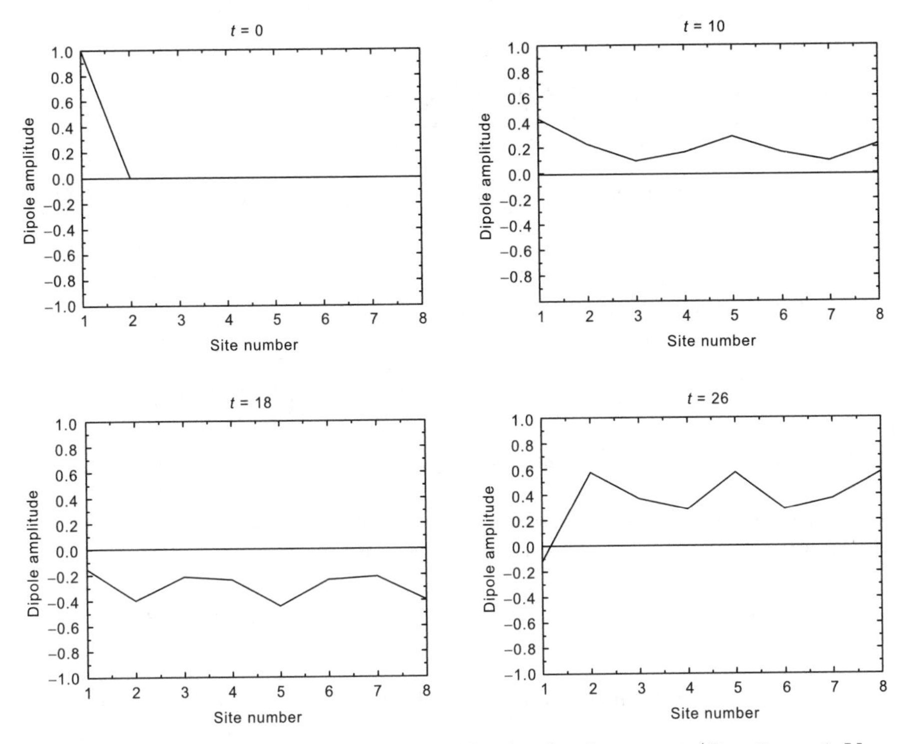

Fig. 14. Time evolution of the dipole distribution for the system ($E = 2, \varepsilon = 1, V = 1$) obtained from the Heisenberg equation with the semiclassical approximation. The initial population differences of sites 4, 5, and 6 are inverted

the terms linear in the coupling constant $g = U/\hbar$ vanish as

$$\tilde{H} = H_0 + \frac{1}{2}[H_{\text{int}}, S] + O(g^3) \ . \tag{50}$$

The transformed Hamiltonian $\tilde{H}$ contains the terms in the second order of the exciton–localized photon interaction that describe exciton–exciton interactions via localized photons. Taking the expectation value of $\tilde{H}$ in terms of the vacuum of localized photons $|vac\rangle$, we obtain the effective Hamiltonian H_{eff} as

$$H_{\text{eff}} = \langle vac|\tilde{H}|vac\rangle = H_b + H_{b-b} \ , \tag{51a}$$

$$H_b = \sum_n \hbar\Omega b_n^\dagger b_n \ , \tag{51b}$$

$$H_{b-b} = \sum_{n,m} \left(\frac{\hbar g^2}{\Omega I - R/\hbar} \right)_{nm} b_n^\dagger b_m \ , \tag{51c}$$

where the exciton energy E is denoted as $\hbar\Omega$. With the help of the components of the Bloch vector (P_n, V_n, W_n) the effective Hamiltonian (51a) can be rewritten as

$$H_{\text{eff}} = \hbar \sum_n (\Omega + \Delta\Omega_n) \frac{1 + W_n}{2} + \frac{\hbar}{4} \sum_n \sum_{m \neq n} \Delta\Omega_{nm} (P_n P_m + V_n V_m) , \quad (52)$$

where the interaction energy or coupling strength between excitons $\Delta\Omega_{nm}$ are given as

$$\Delta\Omega_{nm} = \left(\frac{g^2}{\Omega I - R/\hbar} \right)_{nm} = \sum_j \frac{g^2}{\Omega - \lambda_j} P_{nj} P_{mj} , \quad (53)$$

and the abbreviation $\Delta\Omega_n \equiv \Delta\Omega_{nn}$ is used for a special case.

5.2 Classification of Quasisteady States

Using the effective Hamiltonian H_{eff}, we can obtain the time evolution of the dipole moment at site n in a similar way as discussed in Sect. 4.1:

$$\langle P_n(t) \rangle = \langle P_n \rangle \cos[(\Omega + \Delta\Omega_n)t] + \sum_{m \neq n} \Delta\Omega_{nm} t \langle W_n \rangle \langle P_m \rangle \sin(\Omega t) . \quad (54)$$

If the dipole at site 1, in particular, is only excited at $t = 0$, (54) reads

$$\begin{aligned} \langle P_n(t) \rangle &= \delta_{n1} \langle P_1 \rangle \cos[(\Omega + \Delta\Omega_n)t] \\ &\quad + (1 - \delta_{n1}) \Delta\Omega_{n1} t \langle W_n \rangle \langle P_1 \rangle \sin(\Omega t) . \end{aligned} \quad (55)$$

It follows from this expression that the sign of the dipole at site n depends on the coefficient $\Delta\Omega_{n1}$ as well as the initial values of $\langle P_1(0) \rangle$ and $\langle W_n(0) \rangle$, and that the coefficient $\Delta\Omega_{n1}$ determines what kind of quasisteady states of the dipole distribution the system reaches when the initial values are fixed. Therefore we examine the matrix $\Delta\Omega$ to classify the quasisteady states of the dipole distribution of the system. First, as an example, we show the matrix elements of $\Delta\Omega$ for the system examined in Sect. 3 and Sect. 4 whose material parameters are $E = 2, \varepsilon = 1$, and $V = 1$ as follows:

$$\Delta\Omega = \begin{pmatrix} \frac{1}{3} & \frac{2}{3} & \frac{1}{3} & -\frac{1}{3} & -\frac{2}{3} & -\frac{1}{3} & \frac{1}{3} & \frac{2}{3} \\ \frac{2}{3} & \frac{1}{3} & \frac{2}{3} & \frac{1}{3} & -\frac{1}{3} & -\frac{2}{3} & -\frac{1}{3} & \frac{1}{3} \\ \frac{1}{3} & \frac{2}{3} & \frac{1}{3} & \frac{2}{3} & \frac{1}{3} & -\frac{1}{3} & -\frac{2}{3} & -\frac{1}{3} \\ -\frac{1}{3} & \frac{1}{3} & \frac{2}{3} & \frac{1}{3} & \frac{2}{3} & \frac{1}{3} & -\frac{1}{3} & -\frac{2}{3} \\ -\frac{2}{3} & -\frac{1}{3} & \frac{1}{3} & \frac{2}{3} & \frac{1}{3} & \frac{2}{3} & \frac{1}{3} & -\frac{1}{3} \\ -\frac{1}{3} & -\frac{2}{3} & -\frac{1}{3} & \frac{1}{3} & \frac{2}{3} & \frac{1}{3} & \frac{2}{3} & \frac{1}{3} \\ \frac{1}{3} & -\frac{1}{3} & -\frac{2}{3} & -\frac{1}{3} & \frac{1}{3} & \frac{2}{3} & \frac{1}{3} & \frac{2}{3} \\ \frac{2}{3} & \frac{1}{3} & -\frac{1}{3} & -\frac{2}{3} & -\frac{1}{3} & \frac{1}{3} & \frac{2}{3} & \frac{1}{3} \end{pmatrix} .$$

The matrix is symmetric and has a cyclic structure as

Table 1. Classification of quasisteady states

Group	Signs of $\Delta\Omega_{1n}$ for $n = 1, 2, 3, \cdots, 8$
1st	$(-,-,-,+,+,+,-,-)$, $(+,+,+,-,-,-,+,+)$
2nd	$(-,+,-,-,+,-,-,+)$, $(+,-,+,+,-,+,+,-)$
3rd	$(-,-,-,-,-,-,-,-)$, $(+,+,+,+,+,+,+,+)$
4th	$(-,+,-,+,-,+,-,+)$, $(+,-,+,-,+,-,+,-)$

$$\Delta\Omega_{n+1,m+1} = \Delta\Omega_{n,m} \,.$$

This is due to the fact that the periodic boundary condition is imposed on the system, and that the Hamiltonian given in (4) has translational invariance. Owing to the symmetric structure as

$$(\Delta\Omega_{22}, \Delta\Omega_{23}, \Delta\Omega_{24}, \cdots) = (\Delta\Omega_{11}, \Delta\Omega_{12}, \Delta\Omega_{13}, \cdots) \,,$$

the dipole distribution of the system initiated by the dipole excitation at site 2 can be discussed in terms of $\Delta\Omega_{1n}$ that governs the problem driven by the site-one excitation. In general the problem is reduced to examining the structure of the coefficients $\Delta\Omega_{1n}$ that depend on the material parameters Ω and λ_j, or E, ε, and V. Since the parameter sets crossing the resonant points ($\Omega = \lambda_j$) change the pattern of the dipole distribution in the system, we can classify quasisteady states of the dipole distribution into four groups in Table 1.

The system with $E = 2$, $\varepsilon = 1$, and $V = 1$, which has been investigated in Sect. 4, belongs to the first group. Since the coefficients $\Delta\Omega_{1n}$ in the first group have different signs at sites 4, 5, and 6 from the others, the quasisteady state of the dipole distribution that is predicted by the effective Hamiltonian is the one described by the second-order perturbation solution (see (44) and Fig. 8). The third and fourth groups are, in particular, interesting among the four groups of the quasisteady states. The third group corresponds to a "ferromagnetic" system, in which all the electric dipole moments are aligned in the same direction, while the fourth group corresponds to an "antiferromagnetic" system, where the dipoles are aligned to have alternating signs.

Using numerical results of (46) and (47), we examine the dynamics of the third and fourth groups whether they behave as predicted by the effective Hamiltonian: "ferromagnetic," or "antiferromagnetic." The parameter set of $E = 3.01$, $\varepsilon = 1$, and $V = 1$ gives

$$\Delta\Omega_{1n} = (-13.1, -12.7, -12.4, -12.2, -12.2, -12.2, -12.4, -12.7) \,,$$

and thus the system belongs to the third group. The result is shown in Fig. 15, which predicts the "ferromagnetic" behavior as expected.

When the parameter set of $E = 1$, $\varepsilon = 3.01$, and $V = 1$ is chosen, the system corresponds to the fourth group. Figure 16 presents the time evolution

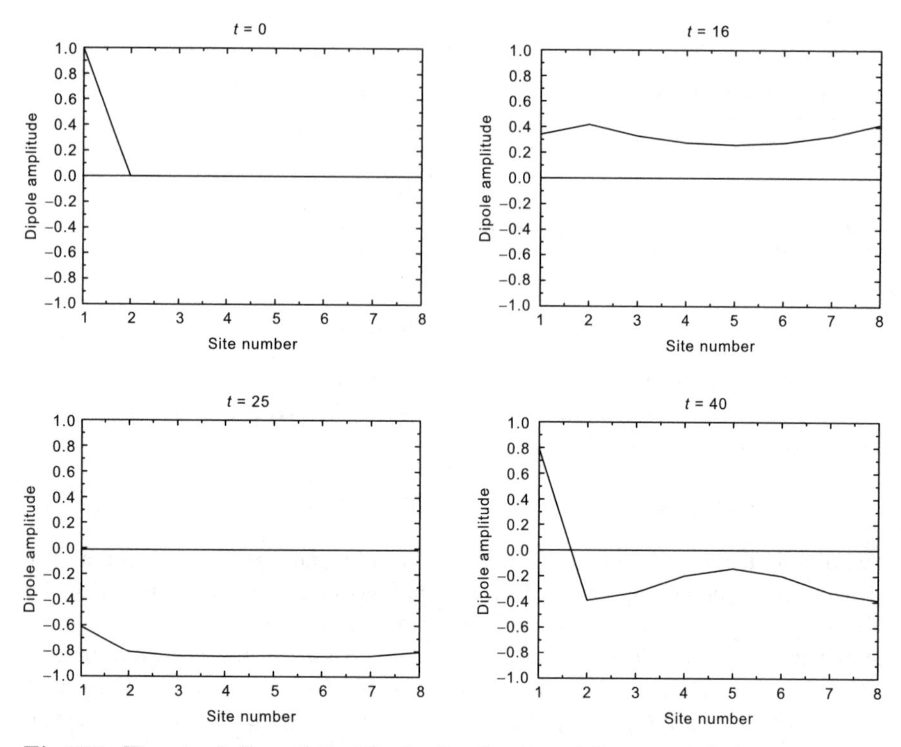

Fig. 15. Time evolution of the dipole distribution of the system belonging to the third group. The initial conditions are set as $\langle P_n \rangle = \delta_{n1}$ and $\langle W_n \rangle = 1 - \delta_{n1}$. All the electric dipole moments are aligned in the same direction, and a dipole-ordered state appears

of the dipole distribution of the system that shows alternating changes in signs, as expected to be similar to antiferromagnetics.

In an even site number "antiferromagnetic" system, the dipole distribution from an initial state evolves to align the dipoles alternatingly up and down, as shown in Fig. 16. On the other hand, an odd site number system becomes frustrated because the dipole distribution cannot be perfectly aligned alternatingly. Thus it is interesting to examine the time evolution of the dipole distribution in an odd site number system. The result for the $N = 9$ "antiferromagnetic" system is presented in Fig. 17. It follows from the figure that there are ordered dipoles near the one initially prepared at site one, and alternating dipoles including a pair at the opposite site to site one. Since strong frustration, for example as occurred in spin glass, should produce a random distribution, the figure indicates that the odd site number system considered here is not frustrated so strongly.

With the help of the flip hypothesis proposed in Sect. 4.2, we infer that the distribution of the dipoles can be changed by individual manipulation of the initial population differences, and that an "antiferromagnetic" state can be

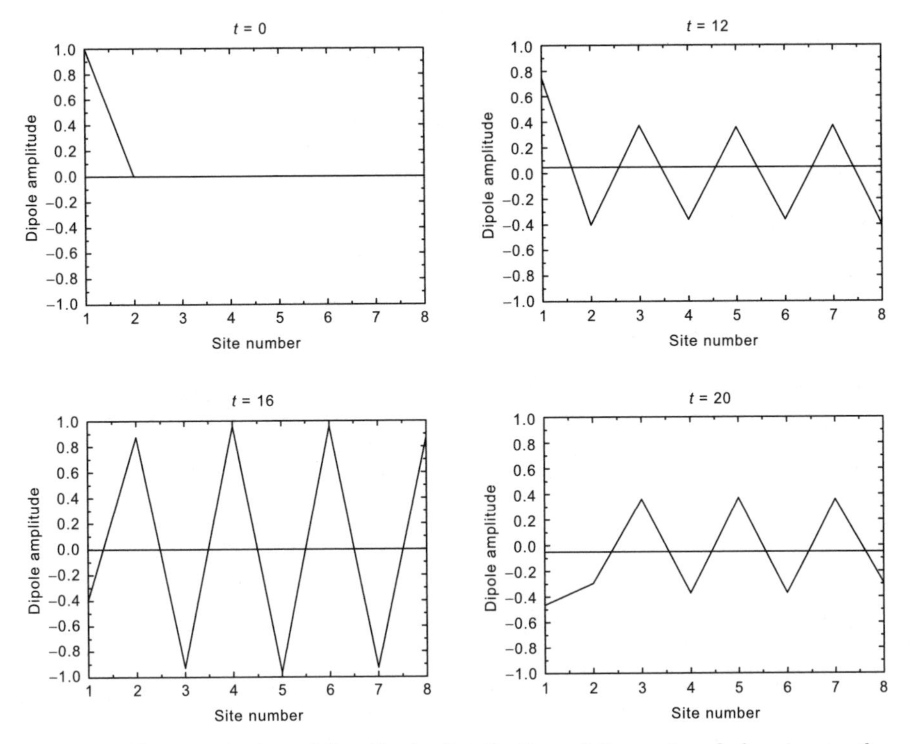

Fig. 16. Time evolution of the dipole distribution of the system belonging to the fourth group. The initial conditions are set as $\langle P_n \rangle = \delta_{n1}$ and $\langle W_n \rangle = 1 - \delta_{n1}$. The dipole distribution similar to the "antiferromagnetic" state is shown, where the dipoles alternatingly change the directions

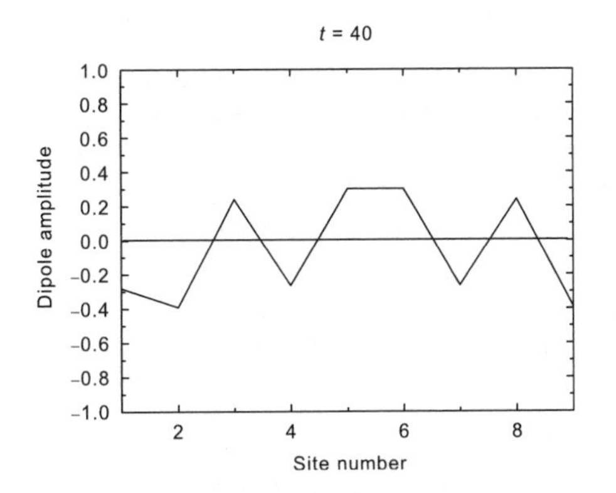

Fig. 17. Quasisteady state of the dipole distribution in the "antiferromagnetic" system belonging to the fourth group. The total number of sites is assumed to be nine

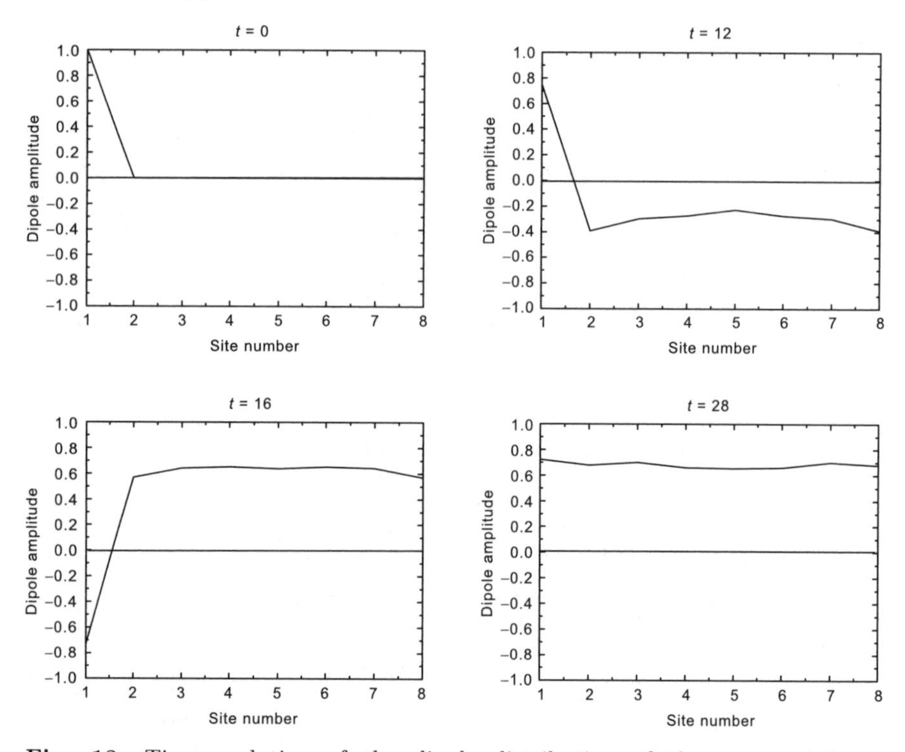

Fig. 18. Time evolution of the dipole distribution of the system belonging to the fourth group. The initial population differences are set as $W_n = (0, -, +, -, +, -, +, -)$. According to the flip hypothesis, the alternating dipoles change their directions and are aligned to form a dipole-ordered state

transformed into a dipole-ordered state. To verify this, the initial population differences are set in the zigzag form as $W_n = (0, -, +, -, +, -, +, -)$ for the "antiferromagnetic" system ($E = 1$, $\varepsilon = 3.01$, and $V = 1$), and the time evolution of the dipole distribution is examined. The result, as shown in Fig. 18, clearly indicates that a dipole-ordered state manifests itself owing to the local manipulation of the initial population differences. It is noteworthy that the amplitude of each dipole is close to one, and that the total dipole of the system is of the order of N, whose radiation property is investigated in the following sections.

5.3 Response to the Initial Input of a Localized Photon: the Robustness of Quasisteady States

Until now, the state of localized photons at each site was initially taken to be a vacuum. Since localized photons in the Heisenberg equations have different degrees of freedom, this section examines the temporal evolution given an initial distribution of localized photons. The initial conditions take the state given below, in which the distribution of localized photons is given for site 1 only.

$$\langle x_n \rangle = x_1 \delta_{n1}, \; \langle y_n \rangle = 0 \;, \quad \langle P_n \rangle = \langle V_n \rangle = 0, \; \langle W_n \rangle = 1 \;. \tag{56}$$

This state can be achieved by having the localized photons at site 1 take on a coherent state in which the eigenvalue is a real number. The coherent state is the eigenstate of the annihilation operator of a simple harmonic oscillator. Therefore, the initial state satisfying the following condition needs to be established at site 1:

$$a_1 |x_1\rangle = x_1 |x_1\rangle \quad x_1 \text{ is real.} \tag{57}$$

The temporal evolution of the dipole distribution is examined by solving the semiclassical Heisenberg equations (46) and (47) numerically with these initial conditions for each of the systems categorized in this section. Figure 19 shows the dynamics of systems in the first group.

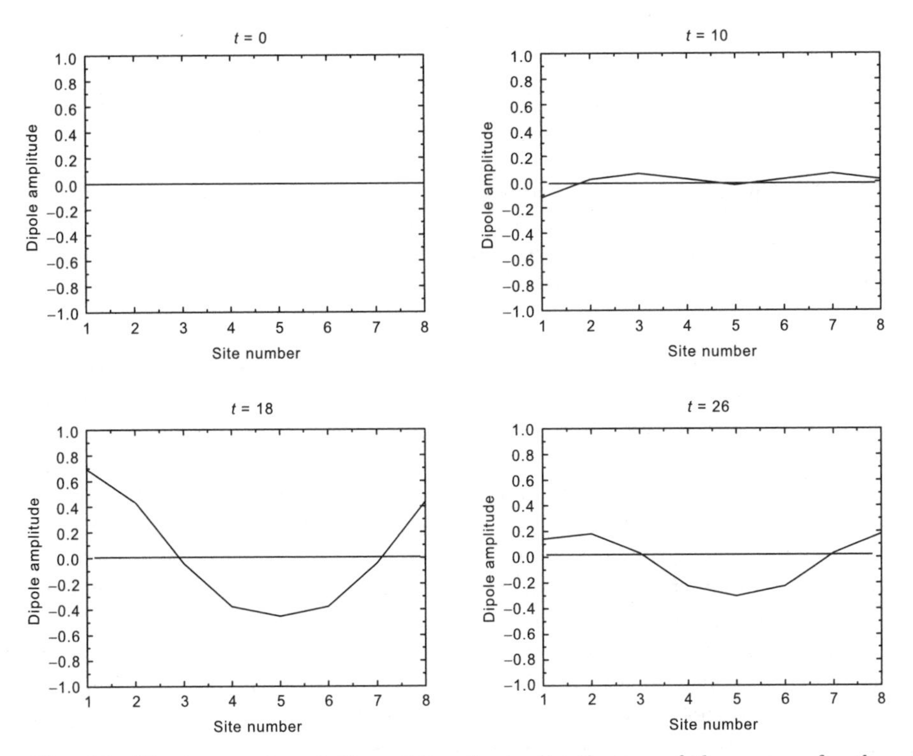

Fig. 19. The temporal evolution of the dipole distribution of the system for the first group in which the localized photonic states are initially in the coherent state at the first site and the other sites are in the vacuum state and all the excitonic states are in the totally inverted state

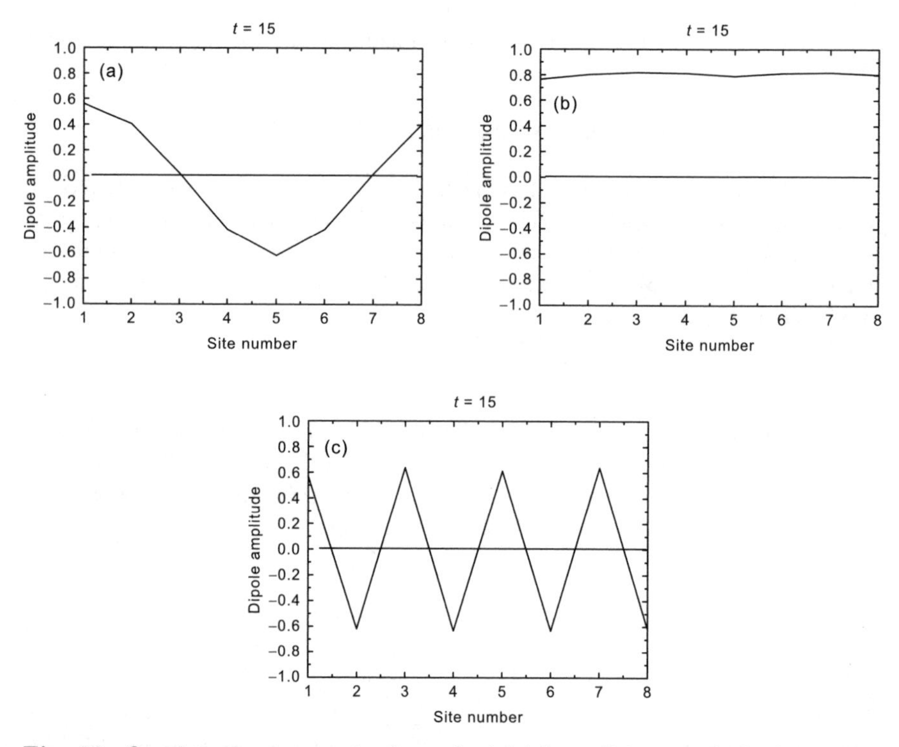

Fig. 20. Quasisteady states arise from the initial condition in which there is localized photon excitation at a single site when the system parameters have various values (ϵ, V, E). The results in (**a**), (**b**), and (**c**) correspond to the first, third, and fourth groups in Table 1, respectively

A quasisteady state emerges that is similar to the case in which dipoles were given as the initial condition (e.g. Fig. 13). Figure 20 shows the quasisteady states corresponding to each of the three systems considered in this section.

These results demonstrate that the states of each group behave exactly as predicted by the effective Hamiltonian. In other words, while the state derived for each group originally involved a dipole (instead of a localized photon), similar states are also realized with the initial input of localized photons. On closer examination of Fig. 19, it can be seen that dipoles are first induced at the site during the initial time frame as a result of the input of localized photons. Therefore, these dipoles established the basis for realizing quasisteady states via a mechanism that is similar to the case in which dipoles were the initial input. These results imply the robustness of each quasisteady state predicted by the effective Hamiltonian against input.

6 Dicke's Superradiance

In the preceding section, we have shown that electric transition dipole moments of the system are aligned to have a large value, depending on the material parameters of the system. It is then expected that strong radiation can be emitted due to a large total dipole moment, and it is intriguing to examine radiation properties of the dipole-ordered states obtained from our localized photon model. Before such an investigation, we will outline superradiance [49, 54–57, 60, 63] as a cooperative phenomenon, where radiation from N two-level systems interacting with propagating light exceeds the sum of the one individually emitted from the systems. More precisely, a radiation pulse with a sharp width of $1/N$ and a high intensity of N^2 is emitted [49]. This kind of system is called the Dicke model, and was extensively studied by the middle of the 1970s since the pioneering work of Dicke in 1954 [49].

In this section we first introduce the Dicke model and Dicke states, and then discuss superradiant states, or superradiance, following Mandel and Wolf [77]. In order to examine how such states are dynamically obtained, a master equation is derived with the help of the projection operator method, and the time evolution of radiation is studied. Numerical solutions of the Dicke master equation are presented for a small system whose size is much smaller than the wavelength of the radiation field. At the same time, analytic solutions are obtained within a semiclassical approximation to discuss the importance of quantum correlation. Finally, we examine the effects of the dipole–dipole interaction for an inhomogeneous system, as well as the effects of reabsorption and re-emission of photons for a large system, both of which violate the cooperative states and destroy some part of the coherence in the system.

6.1 Dicke States and Superradiance

N Two-level Systems and Dicke States

The Dicke model consists of N two-level systems, as a representative of atoms or quantum dots, coupled to the radiation field. The free Hamiltonian for the two-level systems is written as

$$H_{\text{two-level}} = \hbar\Omega \sum_{n=1}^{N} b_n^\dagger b_n \ , \tag{58}$$

where $\hbar\Omega$, $b_n^\dagger$, and b_n denote the excitation energy, creation and annihilation operators of excitation in the n-th site, respectively. These operators can also be defined by spin operators $R_i^{(n)} (i = 1, 2, 3)$ and Bloch vectors (P_n, V_n, W_n) as follows.

$$R_1^{(n)} = \frac{1}{2}(b_n + b_n^\dagger) = \frac{1}{2}P_n \, , \tag{59a}$$

$$R_2^{(n)} = \frac{\mathrm{i}}{2}(b_n - b_n^\dagger) = \frac{1}{2}V_n \, , \tag{59b}$$

$$R_3^{(n)} = \frac{1}{2}(b_n^\dagger b_n - b_n b_n^\dagger) = \frac{1}{2}W_n \, . \tag{59c}$$

The spin operators $R_i^{(n)}$ satisfy the commutation relations

$$[R_i^{(n)}, R_j^{(n')}] = \mathrm{i}\delta_{nn'}\epsilon_{ijk}R_k^{(n)} \, , \tag{60}$$

with Levi-Civita's symbol ϵ_{ijk}. Introducing total spin operators for the N two-level systems as

$$R_1 = \sum_{n=1}^{N} R_1^{(n)} \, , \quad R_2 = \sum_{n=1}^{N} R_2^{(n)} \, , \quad R_3 = \sum_{n=1}^{N} R_3^{(n)} \, , \tag{61}$$

we can rewrite the free Hamiltonian $H_{\text{two-level}}$ as

$$H_{\text{two-level}} = \hbar \Omega R_3 \, , \tag{62}$$

where the total spin operators also obey the commutation relations

$$[R_i, R_j] = \mathrm{i}\epsilon_{ijk}R_k \, , \tag{63}$$

similar to (60).

Let us consider a state in which N_1 of the N two-level systems are in the lower state $|1\rangle$ and N_2 are in the excited state $|2\rangle$, where

$$N = N_1 + N_2 \, , \quad m = \frac{1}{2}(N_2 - N_1) \, . \tag{64}$$

Evidently m is a measure of the total population inversion, and it is an eigenvalue of the spin operator R_3. Introducing the operator $R^2 = R_1^2 + R_2^2 + R_3^2$ and using (63), we obtain the commutation relations

$$[R_i, R^2] = 0 \, , \quad \text{for } i = 1, 2, 3 \, , \tag{65}$$

so that the collective operators R_i obey the same algebra as the angular momentum. Thus there exist the states $|l, m\rangle$ that are simultaneous eigenstates of both R_3 and R^2 as

$$R_3|l, m\rangle = m|l, m\rangle \, , \quad \text{for } \frac{1}{2}N \geq m \geq -\frac{1}{2}N \, , \tag{66a}$$

$$R^2|l, m\rangle = l(l+1)|l, m\rangle \, , \quad \text{for } \frac{1}{2}N \geq l \geq |m| \, . \tag{66b}$$

The quantum number l is called the cooperation number [49] and determines the collectivity of cooperative phenomena. For the raising and lowering operators as

$$R_+ = \sum_{n=1}^{N} b_n^\dagger = R_1 + \mathrm{i}R_2 \ , \quad R_- = \sum_{n=1}^{N} b_n = R_1 - \mathrm{i}R_2 \ , \tag{67}$$

it follows from the well-known properties of angular momentum operators that

$$R_\pm |l, m\rangle = \sqrt{(l \mp m)(l \pm m + 1)} |l, m \pm 1\rangle \ , \tag{68}$$

where the collective states $|l, m\rangle$ are called Dicke states.

Superradiant States

In the following, let us assume that the system size is much smaller than the wavelength of the radiation, and that the long-wave approximation is valid. The interaction Hamiltonian between the two-level system and radiation in the dipole coupling is given by [49]

$$H_{\text{Fint}}(t) = -\boldsymbol{\mu}_{\text{tot}}(t) \cdot \boldsymbol{E}(\boldsymbol{r} = 0, t) \ , \tag{69}$$

where the electric field is evaluated at the origin owing to the long-wave approximation, and the total electric dipole of the system is written as

$$\boldsymbol{\mu}_{\text{tot}} = \boldsymbol{\mu} \sum_{n=1}^{N} (b_n + b_n^\dagger) = \boldsymbol{\mu}(R_+ + R_-) \ . \tag{70}$$

Here we suppose a transition from an initial state $|l, m\rangle|vac\rangle$ to a final state $|\psi\rangle$, where $|l, m\rangle$ and $|vac\rangle$ represent a Dicke state and the vacuum of the radiation field, respectively. Then the transition probability can be written as

$$-\langle\psi|\boldsymbol{\mu} \cdot \boldsymbol{E}_-(t) R_-(t) |l, m\rangle|vac\rangle \ , \tag{71}$$

where $\boldsymbol{E}_-(t)$ is the $\mathrm{e}^{-\mathrm{i}\omega t}$-dependent part of the electric field $\boldsymbol{E}(t)$. By squaring the absolute value of the transition probability and summing over all possible final states, the probability of emission of radiation photons w_{emit} is given as [49, 77]

$$\begin{aligned} w_{\text{emit}} &= \sum_\psi |\langle\psi|\boldsymbol{\mu} \cdot \boldsymbol{E}(t) R_-(t)|l, m\rangle|vac\rangle|^2 \\ &= \langle l, m|R_+(t) R_-(t)|l, m\rangle\langle vac|\boldsymbol{\mu} \cdot \boldsymbol{E}_+(t)\boldsymbol{\mu} \cdot \boldsymbol{E}_-(t)|vac\rangle \\ &= \langle R_+(t) R_-(t)\rangle A = (l + m)(l - m + 1)A \ , \end{aligned} \tag{72}$$

where (68) is used for the final expression, and $A = \langle vac|\boldsymbol{\mu} \cdot \boldsymbol{E}_+(t)\boldsymbol{\mu} \cdot \boldsymbol{E}_-(t)|vac\rangle$ is identified as the Einstein A coefficient for each two-level system like an atom or a quantum dot.

If all quantum dots are in the ground state $|1\rangle$, then $m = -N/2$, and from (66b) we must have $l = N/2$. Hence, from (72) the radiation rate w_{emit}

is zero as expected. If all quantum dots are in the excited states $|2\rangle$, then $m = N/2$ and we again have $l = N/2$. Hence $w_{\mathrm{emit}} = NA$, which is just what one would expect from a group of N independently radiating quantum dots. However, the situation is quite different if the initial state is not the completely excited state. Let us consider a state in which half of the quantum dots are excited and half are not, so that $m = 0$. Then we have

$$w_{\mathrm{emit}} = l(l+1)A \,, \tag{73}$$

and l can have any value between 0 and $N/2$. The larger the value of l, the larger is the collective rate of radiation of the system. In particular, if $l = N/2$, then

$$w_{\mathrm{emit}} = \frac{1}{4}N(N+2)A \,, \tag{74}$$

which maximizes the value of cooperative spontaneous radiation. Since the maximum radiation probability is proportional to the square of the site number N, it becomes very large for large N. This phenomenon is called superradiance, and the state $|l = N/2, m = 0\rangle$ is called a superradiant Dicke state. From the definition of the total Bloch vector (P, V, W) as $P = \sum_n \langle P_n \rangle, V = \sum_n \langle V_n \rangle, W = \sum_n \langle W_n \rangle$, it follows that the Bloch vector for the superradiant state is on the P–V plane and its length reaches a maximum, as shown in Fig. 21.

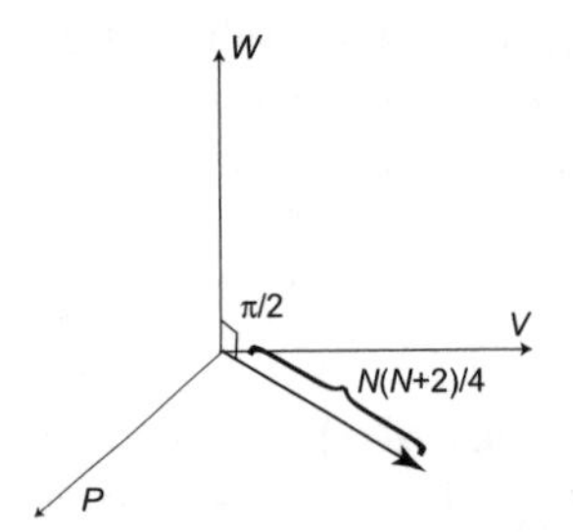

Fig. 21. Total Bloch vector in a superradiant state. It is on the P–V plane, and its length has a maximal value

6.2 Dicke Master Equation and Solutions for a Small System

In this section the Dicke master equation is derived for a small system to discuss dynamical manifestation of superradiant states described in the preceding section.

Nakajima–Zwanzig's Method and Generalized Master Equation

According to Mandel and Wolf [77, 78], we outline the derivation of a generalized master equation by projection operators to project out of the density operator for a total system that part on which our interest is focused [79–82]. The total system can be described in the interaction representation by the Liouville–von Neumann equation for the density operator $\rho(t)$ as

$$\frac{\partial \rho(t)}{\partial t} = \frac{1}{\mathrm{i}\hbar}[H_{\mathrm{I}}(t), \rho(t)] = -\mathrm{i}L(t)\rho(t) , \tag{75}$$

where $H_{\mathrm{I}}(t)$ is the interaction Hamiltonian, and $L\cdots = [H_{\mathrm{I}}(t), \cdots]/\hbar$ is the Liouvillian operator, as before. We now introduce a time-independent projection operator P:

$$P^2 = P , \tag{76}$$

which is chosen so as to project out the most relevant part of $\rho(t)$. Using the complementary operator $(1-P)$, we can express $\rho(t)$ as

$$\rho(t) = P\rho(t) + (1-P)\rho(t) . \tag{77}$$

On multiplying both sides of (75) by P on the left and then using (77) to substitute for $\rho(t)$ on the right, we have

$$P\frac{\partial \rho(t)}{\partial t} = -\mathrm{i}PL(t)P\rho(t) - \mathrm{i}PL(t)(1-P)\rho(t) . \tag{78}$$

Similarly, after multiplying (75) by $(1-P)$ on the left, we obtain

$$(1-P)\frac{\partial \rho(t)}{\partial t} = -\mathrm{i}(1-P)L(t)P\rho(t) - \mathrm{i}(1-P)L(t)(1-P)\rho(t) . \tag{79}$$

Formally integrating the first-order differential equation (79) and solving for $(1-P)\rho(t)$, we find

$$\begin{aligned}(1-P)\rho(t) = {} & \exp\left(-\mathrm{i}(1-P)\int_0^t L(t')\mathrm{d}t'\right)(1-P)\rho(0) \\ & -\mathrm{i}\int_0^t \exp\left(-\mathrm{i}(1-P)\int_\tau^t L(t')\mathrm{d}t'\right) \\ & \times(1-P)L(\tau)P\rho(\tau)\mathrm{d}\tau .\end{aligned} \tag{80}$$

Substitution of $(1-P)\rho(t)$ into (78) gives

$$\begin{aligned}P\frac{\partial \rho(t)}{\partial t} = {} & -\mathrm{i}PL(t)P\rho(t) \\ & -\mathrm{i}PL(t)\exp\left(-\mathrm{i}(1-P)\int_0^t L(t')\mathrm{d}t'\right)(1-P)\rho(0) \\ & -PL(t)\int_0^t \exp\left(-\mathrm{i}(1-P)\int_\tau^t L(t')\mathrm{d}t'\right) \\ & \times(1-P)L(\tau)P\rho(\tau)\mathrm{d}\tau .\end{aligned} \tag{81}$$

If we set

$$U(t,\tau) \equiv \exp\left(-\mathrm{i}(1-P)\int_{\tau}^{t} L(t')\mathrm{d}t'\right) , \tag{82}$$

we finally obtain Nakajima-Zwanzig's generalized master equation [79–82] as follows:

$$\begin{aligned} P\frac{\partial \rho(t)}{\partial t} &= -\mathrm{i}PL(t)P\rho(t) - \mathrm{i}PL(t)U(t,0)(1-P)\rho(0) \\ &\quad -PL(t)\int_0^t U(t,\tau)(1-P)L(\tau)P\rho(\tau)\mathrm{d}\tau . \end{aligned} \tag{83}$$

Application to the Dicke problem

As an example, let us consider the application of (83) to the Dicke problem, i.e., to the interaction of a group of N closely spaced, identical two-level atoms with an electromagnetic field. For simplicity, we assume that two-level atoms are at the origin, and we make the rotating-wave approximation. Then the Hamiltonian is given as [57]

$$H_{\mathrm{D}} = \hbar\Omega R_3 + \hbar\sum_{\boldsymbol{k}\lambda}\omega a_{\boldsymbol{k}\lambda}^{\dagger}a_{\boldsymbol{k}\lambda} + \hbar\sum_{\boldsymbol{k}\lambda}(g_{\boldsymbol{k}\lambda}a_{\boldsymbol{k}\lambda}^{\dagger}R_{-} + h.c.) , \tag{84}$$

where $a_{\boldsymbol{k}\lambda}^{\dagger}(a_{\boldsymbol{k}\lambda})$ is the photon creation (annihilation) operator specified by wavevector $\boldsymbol{k}$, polarization λ, and frequency ω, and the coupling constant $g_{\boldsymbol{k}\lambda}$ is given by the formula

$$g_{\boldsymbol{k}\lambda} = \frac{\mathrm{i}\Omega}{\sqrt{2\epsilon_0\hbar\omega V}}(\boldsymbol{\epsilon}_{\boldsymbol{k}\lambda}\cdot\boldsymbol{\mu}) . \tag{85}$$

Here $\boldsymbol{\epsilon}_{\boldsymbol{k}\lambda}$ represents the polarization vector, and $\boldsymbol{\mu}$ denotes the matrix element of the dipole moment in the two-level (atomic) system.

In the following, we are mainly interested in the two-level (atomic) system as the relevant system and identify the 'reservoir' with the radiation field, so that the projection operator P is given by

$$P\cdots = |vac\rangle\langle vac|\mathrm{Tr}_{\mathrm{F}}\cdots , \tag{86}$$

where Tr_{F} denotes that the trace is taken over the radiation field states. As the initial state of the coupled system, we also assume the direct product state with the radiation field in the vacuum state

$$\rho(0) = \rho_A(0)\otimes|vac\rangle\langle vac| , \tag{87}$$

with the reduced density operator defined as $\rho_A(t) = \mathrm{Tr}_{\mathrm{F}}\rho(t)$. In order to work in the interaction picture, we transform the interaction Hamiltonian into

$$H_{\mathrm{Fint}}(t) = \hbar\sum_{\boldsymbol{k}\lambda}(g_{\boldsymbol{k}\lambda}a_{\boldsymbol{k}\lambda}^{\dagger}R_{-}\mathrm{e}^{\mathrm{i}(\omega-\Omega)t} + h.c.) . \tag{88}$$

Under these conditions, the first term on the right-hand side of (83), $PL(t)P\rho(t)$, can be written as

$$
\begin{aligned}
PL(t)P\rho(t) &= |vac\rangle\langle vac|\mathrm{Tr_F}\frac{1}{\hbar}[H_{\mathrm{Fint}}(t), |vac\rangle\langle vac|\rho_A(t)] \\
&= |vac\rangle\langle vac|\left(\sum_{\boldsymbol{k}\lambda} g_{\boldsymbol{k}\lambda}\langle vac|a^{\dagger}_{\boldsymbol{k}\lambda}|vac\rangle R_{-}\mathrm{e}^{\mathrm{i}(\omega-\Omega)t}\rho_A(t) - h.c.\right) \\
&= 0\ ,
\end{aligned}
\tag{89}
$$

because

$$
\langle vac|a^{\dagger}_{\boldsymbol{k}\lambda}|vac\rangle = 0 = \langle vac|a_{\boldsymbol{k}\lambda}|vac\rangle\ . \tag{90}
$$

In addition, the second term on the right-hand side of (83) can be rewritten to

$$
(1-P)\rho(0) = (1-|vac\rangle\langle vac|\mathrm{Tr_F})\rho_A(0)|vac\rangle\langle vac| = 0\ . \tag{91}
$$

Substituting (89) and (91) into (83), we have

$$
P\frac{\partial\rho}{\partial t} = -PL(t)\int_0^t L(\tau)P\rho(\tau)\mathrm{d}\tau\ , \tag{92}
$$

where we approximate $U(t,\tau)\sim 1$ (Born approximation). Explicit use of (86) yields the equation

$$
\frac{\partial\rho_A}{\partial t} = -\mathrm{Tr_F}L(t)\int_0^t L(\tau)|vac\rangle\langle vac|\rho_A(\tau)\mathrm{d}\tau\ . \tag{93}
$$

After transforming the variable as $\tau = t-\tau'$ in the last integral, we can reduce (93) to

$$
\frac{\partial\rho_A}{\partial t} = -\mathrm{Tr_F}L(t)\int_0^t L(t-\tau')|vac\rangle\langle vac|\rho_A(t-\tau')\mathrm{d}\tau'\ . \tag{94}
$$

Further approximation of replacing $\rho_A(t-\tau')$ by $\rho_A(t)$, which is known as the Markov approximation, provides the final relation

$$
\begin{aligned}
\frac{\partial\rho_A}{\partial t} &= -\mathrm{Tr_F}L(t)\int_0^t L(t-\tau)|vac\rangle\langle vac|\rho_A(t)\mathrm{d}\tau \\
&= -\frac{1}{\hbar^2}\mathrm{Tr_F}\int_0^t [H_{\mathrm{Fint}}(t),[H_{\mathrm{Fint}}(t-\tau),|vac\rangle\langle vac|\rho_A(t)]]\mathrm{d}\tau \\
&= -\frac{1}{\hbar^2}\mathrm{Tr_F}\int_0^t \{H_{\mathrm{Fint}}(t)H_{\mathrm{Fint}}(t-\tau)|vac\rangle\langle vac|\rho_A(t) \\
&\quad -H_{\mathrm{Fint}}(t)|vac\rangle\langle vac|\rho_A(t)H_{\mathrm{Fint}}(t-\tau)\}\mathrm{d}\tau + h.c.
\end{aligned}
\tag{95}
$$

Noticing (88) and the relations

$$\langle vac|a_{\boldsymbol{k}\lambda}a^{\dagger}_{\boldsymbol{k}'\lambda'}|vac\rangle = \delta_{\boldsymbol{k}\boldsymbol{k}'}\delta_{\lambda\lambda'} \, ,$$
$$\langle vac|a^{\dagger}_{\boldsymbol{k}\lambda}a^{\dagger}_{\boldsymbol{k}'\lambda'}|vac\rangle = \langle vac|a_{\boldsymbol{k}\lambda}a_{\boldsymbol{k}'\lambda'}|vac\rangle = \langle vac|a^{\dagger}_{\boldsymbol{k}\lambda}a_{\boldsymbol{k}'\lambda'}|vac\rangle = 0 \, , \quad (96)$$

we have the following equation

$$\frac{\partial \rho_A}{\partial t} = -\int_0^t \sum_{\boldsymbol{k}\lambda} |g_{\boldsymbol{k}\lambda}|^2 \{ \mathrm{e}^{\mathrm{i}(\Omega-\omega)\tau} R_+ R_- \rho_A(t) + h.c. \\ -2\cos(\omega-\Omega)\tau R_- \rho_A(t) R_+ \} \mathrm{d}\tau \, . \quad (97)$$

In the limit $t \gg 1/\Omega$ we put

$$\int_0^t \sum_{\boldsymbol{k}\lambda} |g_{\boldsymbol{k}\lambda}|^2 \mathrm{e}^{\mathrm{i}(\Omega-\omega)\tau} \mathrm{d}\tau \equiv \beta + \mathrm{i}\gamma \, , \quad (98)$$

where β is half the Einstein A coefficient and γ is a frequency shift due to the interaction. Finally we obtain the appropriate master equation describing spontaneous emission from a collective two-level (atomic) system, which is called the Dicke master equation [56, 57, 59]

$$\frac{\partial \rho_A}{\partial t} = -\beta(R_+R_-\rho_A - 2R_-\rho_A R_+ + \rho_A R_+ R_-) - \mathrm{i}\gamma[R_+R_-, \rho_A(t)] \, , \quad (99)$$

or

$$\frac{\partial \rho_A}{\partial t} = -\beta([R_+, R_-\rho_A] + [\rho_A R_+, R_-]) - \mathrm{i}\gamma[R_+R_-, \rho_A(t)] \, . \quad (100)$$

Numerical Solutions of Dicke Master Equation

The rate of photon emission was obtained perturbatively in Sect. 6.1. In this section, we examine the time evolution of this emission rate, using the Dicke master equation (99). Since the rate of energy dissipation from the two-level (atomic) system is proportional to the emission rate, $-\partial\langle R_3\rangle/\partial t$ should be evaluated. Multiplying (99) by R_3 on the right and taking the trace, one finds

$$\frac{\partial}{\partial t}\langle R_3\rangle = \beta\{-\langle R_3R_+R_-\rangle + 2\langle R_+R_3R_-\rangle - \langle R_+R_-R_3\rangle\} \\ -\mathrm{i}\gamma\{\langle R_3R_+R_-\rangle - \langle R_+R_-R_3\rangle\} \, . \quad (101)$$

In order to simplify the above equation, we use the following relations

$$[R_+, R_3] = -R_+ \, , \quad (102)$$
$$R_+R_- = R_3 + \frac{1}{2}N\sum_n \sum_{m\neq n} b^{\dagger}_n b_m \, , \quad (103)$$

which follow from the collective operators defined in (67) and the commutation relations in (63). In addition, using (103), we can show that

$$[R_+R_-, R_3] = \sum_n \sum_{m \neq n} [b_n^\dagger b_m, R_3]$$
$$= \sum_n \sum_{m \neq n} [b_n^\dagger b_m - b_n^\dagger b_m] = 0 \,, \tag{104}$$

and rewrite (101) as

$$\frac{\partial}{\partial t}\langle R_3 \rangle = 2\beta[\langle R_+R_3R_- \rangle - \langle R_3R_+R_- \rangle] \,. \tag{105}$$

With the help of (102), this leads to the final equation

$$\frac{\partial}{\partial t}\langle R_3 \rangle = 2\beta[\langle (R_3R_+ - R_+)R_- \rangle - \langle R_3R_+R_- \rangle]$$
$$= -2\beta\langle R_+R_- \rangle \,. \tag{106}$$

It follows from the above equation that the time evolution of $\langle R_+R_- \rangle$ is required to evaluate the rate of energy dissipation. Multiplying (99) by R_+R_- on the right and taking the trace, we then find

$$\frac{\partial}{\partial t}\langle R_+R_- \rangle = -2\beta\{\langle R_+R_-R_+R_- \rangle - \langle R_+R_+R_-R_- \rangle\}$$
$$-\mathrm{i}\gamma\langle [R_+R_-, R_+R_-] \rangle$$
$$= -4\beta\{\langle R_+R_- \rangle - \langle R_+R_-R_3 \rangle\} \,, \tag{107}$$

where we used the following relation to obtain the second line from the first line

$$R_+R_+R_-R_- = R_+R_-R_+R_- - 2R_+R_- + 2R_+R_-R_3 \,. \tag{108}$$

Neglecting the quantum correlation and putting

$$\langle R_+R_-R_3 \rangle = \langle R_+R_- \rangle \langle R_3 \rangle \,, \tag{109}$$

we then have

$$\frac{\partial}{\partial t}\langle R_+R_- \rangle = -2\beta\langle R_+R_- \rangle(1 - \langle R_3 \rangle) \,. \tag{110}$$

Simultaneously solving (106) and (110), one can obtain the final result. Or putting $z = \langle R_3 \rangle$, this leads to the following nonlinear equation

$$\ddot{z} = -4\beta\dot{z}(1 - z) \,. \tag{111}$$

This allows us to evaluate the rate of energy dissipation, $-\partial \langle R_3 \rangle / \partial t = -\dot{z}$, and thus to obtain the rate of photon emission.

Under the following initial conditions

$$\langle R_3 \rangle(0) = N/2 \,, \quad \langle R_+R_- \rangle(0) = N \,, \tag{112}$$

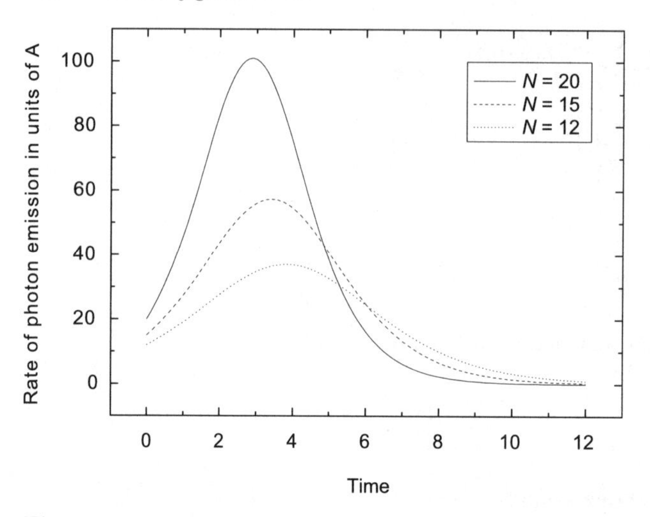

Fig. 22. Time evolution of the rate of photon emission that is numerically obtained from (111) with the initial condition of a totally inverted state. The vertical axis is represented in units of the Einstein A coefficient $A = 2\beta$ for $\beta = 0.05$. The *dotted*, *dashed*, and *solid curves* show the results for the site number $N = 12$, 15, and 20, respectively

which correspond to $l(0) = N/2$ and $m(0) = N/2$ in terms of Dicke states, we solve (111) numerically. The result is shown in Fig. 22 for the case of $\beta = 0.05$ and the site number $N = 12, 15$, and 20. It follows from Fig. 22 that the peak value of the rate of emission becomes higher and the width gets narrower, as the site number becomes larger. As will be shown in the next subsection, a pulse of radiation whose height is proportional to N^2 and width is proportional to $1/N$ is emitted from the two-level system prepared in the initial condition.

Analytical Solutions of Dicke Master Equation in a Semiclassical Approximation

In the preceding subsection, we obtained simultaneously coupled equations after neglecting the quantum correlation more than third order, for example, approximating $\langle R_+R_-R_3\rangle$ as $\langle R_+R_-\rangle\langle R_3\rangle$, and numerically examined the emission rate of radiation to predict a characteristic feature. Here in this section, an analytical solution will be given for time evolution of the emission rate after making more approximations [55, 58, 59].

Using the polar angle θ of a Bloch vector (P, V, W), we can express the expectation value of R_3 as

$$\langle R_3\rangle = \frac{N}{2}\cos\theta \tag{113}$$

and that of R_+R_- as

$$\langle R_+R_-\rangle = \langle R_+\rangle\langle R_-\rangle = \langle R_1\rangle^2 + \langle R_2\rangle^2$$
$$= \langle R\rangle^2 - \langle R_3\rangle^2 = \frac{N^2}{4}\sin^2\theta \,, \tag{114}$$

where we neglected the quantum correlations more than second order as $\langle R_+R_-\rangle = \langle R_+\rangle\langle R_-\rangle$, which is called a semiclassical approximation [57]. Substitution of $z = \cos\theta$ into (106) leads to

$$\dot{z} = -A\frac{N}{2}(1 - z^2) \,, \tag{115}$$

where A is the Einstein A coefficient and is set to $A = 2\beta$. Similarly we can rewrite the left-hand side of (110) as

$$\frac{\partial}{\partial t}\frac{N^2}{4}(1 - z^2) = -\frac{N^2}{2}z\dot{z} \,, \tag{116}$$

and the right-hand side as

$$-2A\frac{N^2}{4}(1 - z^2)\left(1 - \frac{N}{2}z\right) \simeq A\frac{N^3}{4}(1 - z^2)z \,, \tag{117}$$

where we assumed $N \gg 1$. Both results allow us to simplify Eq. (110), and one finds

$$z\dot{z} = -A\frac{N}{2}z(1 - z^2) \,, \tag{118}$$

which is the same as (115). Thus, both (106) and (110) can be transformed to the same equation,

$$\frac{1}{\Gamma}\frac{\mathrm{d}z}{1 - z^2} = -\mathrm{d}t \,, \tag{119}$$

with $\Gamma = AN/2$, and can be easily solved as

$$z = -\tanh\Gamma(t - t_0) \,. \tag{120}$$

We can finally obtain the analytical solution of the emission probability as

$$I \propto -\partial_t\langle R_3\rangle = -\frac{N}{2}\dot{z} = \frac{\Gamma^2}{A}\mathrm{sech}^2\Gamma(t - t_0) \,, \tag{121}$$

which represents a cowbell-shaped function with a width of $2.6/N\beta$ and a peak position at $t = t_0$. This function is plotted in Fig. 23 for the case of $\beta = 0.05$, $t_0 = 3$, and $N = 20$. Comparing the result for $N = 20$ in Fig. 22 with the current result, one finds that a semiclassical treatment can describe radiation profile of the Dicke model very well. However, it should be noted that there must be nonzero total dipole moment in the initial condition in the semiclassical description because the semiclassical approximation is not valid for the totally inverted system [58, 59], where quantum correlation should be essentially included.

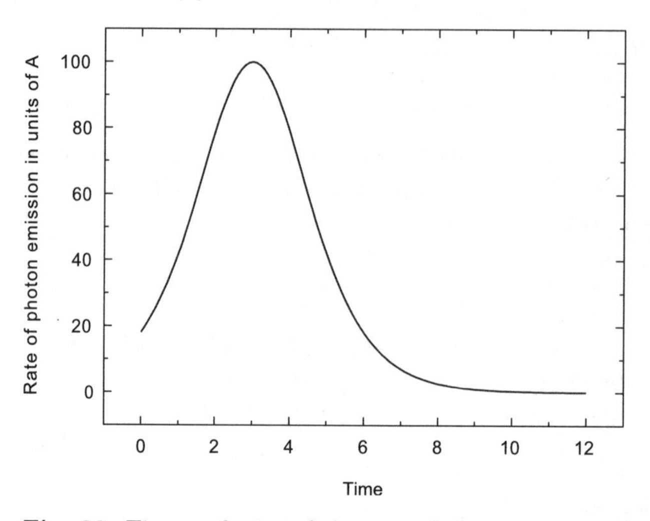

Fig. 23. Time evolution of the rate of photon emission plotted according to the analytic solution obtained with the semiclassical approximation. The vertical axis is represented in units of the Einstein A coefficient $A = 2\beta$. The parameters are set to $\beta = 0.05$, $t_0 = 3$, and $N = 20$, corresponding to Fig. 22

6.3 Effect of the Dipole–Dipole Interaction

The dipole–dipole interaction generally breaks the permutation symmetry of the atom (two-level system)–field coupling because various atoms in the sample have different close-neighbor environments. This perturbation is analogous to an inhomogeneous dephasing of the dipoles, leading to a loss of dipole–dipole correlation. In order to discuss such an effect, we employ the following interaction Hamiltonian, instead of (69)

$$H_{\mathrm{Fint}} = -\sum_{n=1}^{N} \{\boldsymbol{E}^{+}(\boldsymbol{r}_n) \cdot \boldsymbol{P}(\boldsymbol{r}_n) + \boldsymbol{E}^{-}(\boldsymbol{r}_n) \cdot \boldsymbol{P}^{\dagger}(\boldsymbol{r}_n)\} , \tag{122}$$

where the electric field operator $\boldsymbol{E}$ is divided into the positive and negative frequency parts $\boldsymbol{E}^{+}$ and $\boldsymbol{E}^{-}$ defined as

$$\boldsymbol{E}^{+}(\boldsymbol{r}_n) = \mathrm{i} \sum_{\boldsymbol{k}\lambda} \sqrt{\frac{\hbar c k}{2\epsilon_0 V}} \boldsymbol{e}_{\lambda} a_{\boldsymbol{k}\lambda} \mathrm{e}^{\mathrm{i}\boldsymbol{k}\cdot\boldsymbol{r}_n} , \tag{123a}$$

$$\boldsymbol{E}^{-}(\boldsymbol{r}_n) = (\boldsymbol{E}^{+}(\boldsymbol{r}_n))^{\dagger} . \tag{123b}$$

The creation and annihilation operators $a^{\dagger}_{\boldsymbol{k}\lambda}$ and $a_{\boldsymbol{k}\lambda}$ act on the mode of the radiation field with wavevector $\boldsymbol{k}$ and polarization $\boldsymbol{e}_{\lambda}$, and obey the boson commutation relation. An arbitrary quantization volume is denoted as V. The dipole operator is similarly defined as

$$\boldsymbol{P}(\boldsymbol{r}_n) = \mu \boldsymbol{\epsilon} (b^{\dagger}_n + b_n) . \tag{124}$$

Using this interaction Hamiltonian, we can obtain the following master equation in the Born–Markov approximation [62, 83]

$$\frac{\mathrm{d}\rho}{\mathrm{d}t} = -\mathrm{i}\sum_{n\neq m} \Omega_{nm}[b_n^\dagger b_m, \rho] - \sum_{nm} \gamma_{nm}[b_n^\dagger b_m \rho + \rho b_n^\dagger b_m - 2b_m \rho b_n^\dagger] , \quad (125)$$

with

$$\gamma_{nm} = \frac{\mu^2 \Omega^3}{16\pi^2 c^3 \epsilon_0 \hbar} F(\Omega r_{nm}/c) , \quad (126a)$$

$$\Omega_{nm} = -\frac{\mu^2}{16\pi^3 \epsilon_0} \int_0^\infty \mathrm{d}k k^3 F(kr_{nm}) \frac{P}{k - \Omega/c} , \quad (126b)$$

$$F(kr_{nm}) = \int \mathrm{d}\Omega_k (\boldsymbol{e}\cdot\boldsymbol{\epsilon})^2 \mathrm{e}^{\mathrm{i}\boldsymbol{k}\cdot\boldsymbol{r}_{nm}} , \quad (126c)$$

where Ω is the angular frequency corresponding to the excitation energy of the two-level system, and r_{mn} is the absolute value of the vector $\boldsymbol{r}_{nm} = \boldsymbol{r}_n - \boldsymbol{r}_m$. The solid angle Ω_k is defined as $\mathrm{d}\boldsymbol{k} = k^2 \mathrm{d}k \mathrm{d}\Omega_k$. Moreover, we have used the following formula

$$\int_0^\infty c\mathrm{d}\tau \mathrm{e}^{\pm\mathrm{i}(kc-\Omega)\tau} = \pi\delta(k - \Omega/c) \pm \mathrm{i}\frac{P}{k - \Omega/c} , \quad (127)$$

whose real and imaginary parts produce site-dependent relaxation coefficient γ_{nm} and frequency shift Ω_{nm}, respectively. The integral in (126c) can be explicitly calculated and gives [54]

$$\begin{aligned} F(kr_{nm}) &= 4\pi \left(1 - \frac{(\boldsymbol{\epsilon}\cdot\boldsymbol{r}_{nm})^2}{r_{nm}^2}\right) \frac{\sin kr_{nm}}{kr_{nm}} \\ &\quad + 4\pi \left(1 - 3\frac{(\boldsymbol{\epsilon}\cdot\boldsymbol{r}_{nm})^2}{r_{nm}^2}\right) \left(\frac{\cos kr_{nm}}{(kr_{nm})^2} - \frac{\sin kr_{nm}}{(kr_{nm})^3}\right) , \end{aligned} \quad (128)$$

and one readily obtains, at the limit $kr_{nm} \to 0$

$$\hbar\Omega_{nm} = \frac{\mu^2}{4\pi\epsilon_0 r_{nm}^3} \left(1 - 3\frac{(\boldsymbol{\epsilon}\cdot\boldsymbol{r}_{nm})^2}{r_{nm}^2}\right) , \quad (129)$$

which is the static, or Coulombic dipole–dipole interaction.

In order to examine the effects of the dipole–dipole interactions, we evaluate the magnitude of the total Bloch vector $\langle R^2 \rangle$, which is conserved in Dicke states. According to (125), one finds [62]

$$\begin{aligned} \frac{\mathrm{d}}{\mathrm{d}t}\langle R^2 \rangle &= -\mathrm{i}\left\langle [R^2, \sum_{n\neq m} \Omega_{nm} b_n^\dagger b_m] \right\rangle \\ &= -2\mathrm{i} \sum_{n\neq m\neq l} \Omega_{nm} \langle R_z^{(m)} (b_n^\dagger b_l - b_l^\dagger b_n) \rangle , \end{aligned} \quad (130)$$

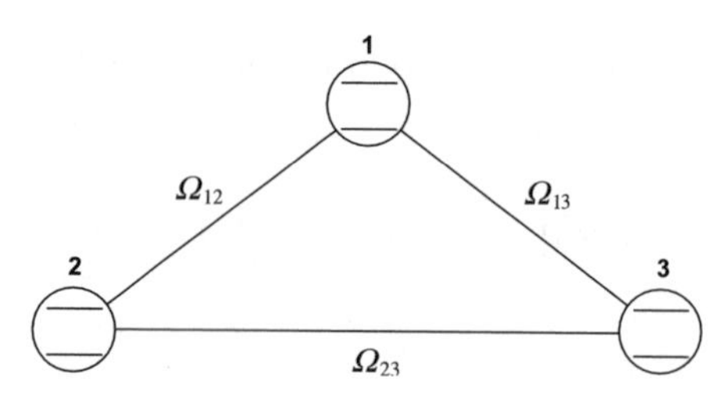

Fig. 24. Arrangement of three two-level systems. Each two-level system is configured at the apex of an isosceles triangle

from which it follows that $\langle R^2 \rangle$ is generally not conserved, but is a conserved quantity only if the site number N is equal to 2, or only if Ω_{nm} is constant for all suffixes [62, 69]. Therefore the state is not expressed by Dicke states, and quantum coherence of the two-level system is destroyed by the dipole–dipole interactions.

Now we consider how the dipole–dipole interactions affect superradiance. Let the site number be three and be arranged as shown in Fig. 24. For simplicity, we assume $\Omega_{12} = \Omega_{13} \neq \Omega_{23}$ and $\gamma_{mn} = \gamma$.

The following eight bases are employed to take matrix elements of (125) and to solve the simultaneous differential equations.

$$|s_1, s_2, s_3\rangle \quad (s_i = \pm) , \tag{131}$$

where the sign + indicates that the upper level is occupied, while the sign − means that the lower level is occupied. Under the initial condition of $\langle + + +|\rho(0)| + ++\rangle = 1$, we obtain the matrix elements of the density operator $\rho(t)$ at time t to express the radiation intensity

$$I(t) = \sum_{n,m} \langle b_n^\dagger b_m \rangle = \langle R_+ R_- \rangle . \tag{132}$$

Using the Pauli matrices σ_μ ($\mu = 0, 1, 2, 3$), we can write nonzero components of the density matrices as $\sum_\mu x_\mu \sigma_\mu$. The vector $\boldsymbol{x} = (x_1, x_2, x_3)$ executes fast precession about a "torque" $\boldsymbol{\Gamma} = \left(2\sqrt{2}\Omega_{12}, 0, \Omega_{23}\right)$ (see Fig. 25). When we introduce the rotation $R\left(\beta, \hat{2}\right)$ that rotates about the 2 axis through an angle

$$\tan\beta = \frac{2\sqrt{2}\Omega_{12}}{\Omega_{23}} , \tag{133}$$

then the vector $\boldsymbol{\Gamma}$ is transformed as $R\left(\beta, \hat{2}\right) \boldsymbol{\Gamma} = (0, 0, \Gamma)$ with

$$\Gamma = \sqrt{\left(2\sqrt{2}\Omega_{12}\right)^2 + \Omega_{23}^2} .$$

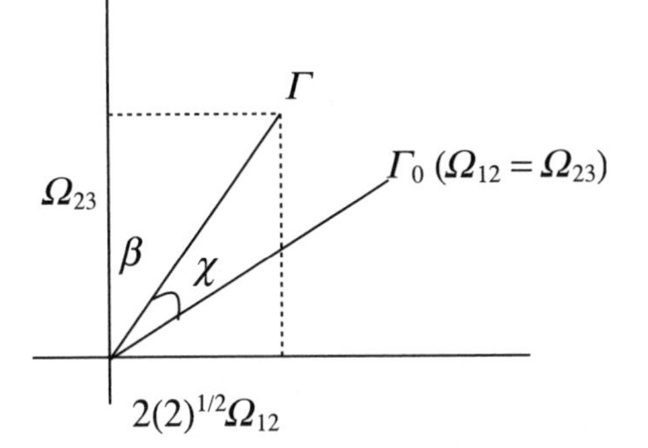

Fig. 25. Relation between the angles β and χ. In the case of pure superradiance ($\Omega_{12} = \Omega_{23}$), the torque Γ is given by Γ_0

The solutions of the differential equations for the density matrices can be obtained from a similar rotation of $\boldsymbol{x}$ with $\cos\chi$ defined as [62]

$$\cos(\chi + \beta) = \frac{1}{3} . \tag{134}$$

When $\cos\chi = 1$ or $\chi = 0$, $\Omega_{12} = \Omega_{23}$ is satisfied and the Dicke superradiant states become exact eigenstates, while $\cos\chi = 0$ gives orthogonal states to the superradiant states. This means that the deviation of χ from zero shows the inhomogeneity of the system, or a measure of the coherence properties.

According to Coffey and Friedberg [62], we present the radiation intensity $I(t)$ in Fig. 26, which shows the peak and tail of $I(t)$ become reduced and longer, respectively, as $\cos\chi$ varies from one to zero corresponding to the sequence of a, c, d, and b in Fig. 26.

Tokihiro et al. [69] also examined the effects of the dipole–dipole interaction on superradiance, using a linear excitonic system, in which excitons can hop from one site to its nearest neighbors via the dipole–dipole interaction. They used the following master equation in the interaction representation

$$\begin{aligned}\frac{\mathrm{d}\rho}{\mathrm{d}t} &= -\mathrm{i}J\sum_n [b_n^\dagger b_{n+1} + b_{n+1}^\dagger b_n, \rho] \\ &\quad -\gamma[R_+R_-\rho + \rho R_+R_- - 2R_-\rho R_+] ,\end{aligned} \tag{135}$$

where the hopping energy of an exciton and the radiation decay rate are denoted as

$$2\hbar J = \frac{\mu^2}{4\pi\epsilon_0 a^3}\left(1 - 3\frac{(\boldsymbol{\epsilon}\cdot\boldsymbol{a})^2}{a^2}\right) , \tag{136}$$

$$\gamma = \frac{\mu^2\Omega^3}{3\pi c^3\epsilon_0\hbar} , \tag{137}$$

respectively. The first term in (135) corresponds to the Hamiltonian of the XY-model in a spin system whose exact solutions are well known [84]. Tak-

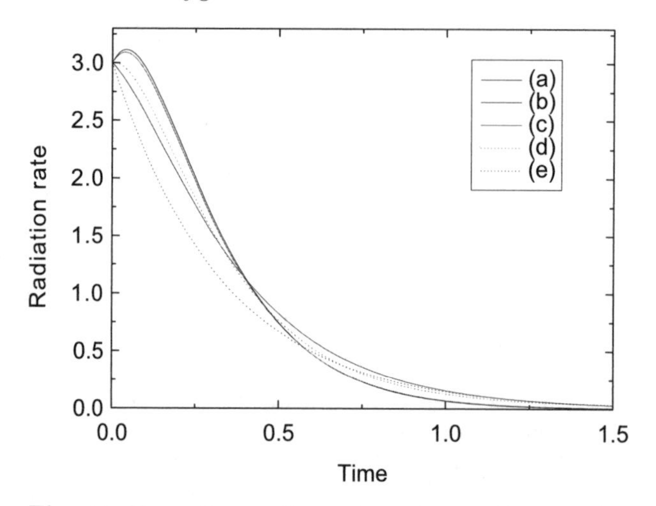

Fig. 26. Normalized radiation intensity for three two-level systems with different Ω_{12}/Ω_{23}: (**a**) $\Omega_{12} = \Omega_{23}$, $\cos\chi = 1$, corresponding to pure superradiance, (**b**) $\cos\chi = 0$, (**c**) $\Omega_{12} = 8\Omega_{23}$, $\cos\chi = 0.96$, (**d**) $\Omega_{12} = \Omega_{23}/8$, $\cos\chi = 0.63$, (**e**) $I = 3\mathrm{e}^{-\gamma t}$, where three two-level systems radiate independently. The vertical axis is represented in units of the Einstein A coefficient while the horizontal axis is represented by γt

ing matrix elements of (135) in terms of the solutions as basis functions, they numerically solved the master equation under the initial condition of a totally inverted system. The result, similar to that of Coffey and Friedberg [62], shows a reduced peak intensity with extension of the tail, and a substantial deviation from Dicke's superradiance that is proportional to the system size N. In addition, they examined the initial-condition dependence, and showed the oscillation of the radiation profile obtained from a partially excited state [83]. This resembles the effects of reabsorption and re-emission of photons, which is discussed in the next subsection.

6.4 Large-Sample Superradiance

In the preceding section, we discussed superradiance in a small system whose size is much smaller than the wavelength of the radiation field. Now we consider large-sample superradiance, taking into consideration a finite system size. In such a system, an inhomogeneous arrangement in the two-level system cannot be neglected, as well as the effects of reabsorption of emitted photons. Dipole–dipole interactions, in particular, become important for the inhomogeneous system and correlation of dipoles in the system is degraded to result in weakened radiation intensity [62, 69], as discussed in the preceding subsection. Reabsorption of emitted photons causes stimulated absorption and emission of photons in the system, which leads to oscillation of radiated pulses [63, 65]. We will analyze this effect semiclassically, according to Banifacio et al. [60, 63, 85], and comment on experimental observations.

Effect of Reabsorption and Reemission of Photons

As samples have a volume with linear dimensions much larger than the wavelength of the radiation, some quantitative differences appear in the dynamics of the system evolution due to the effects connected with the propagation of the electromagnetic field along the emitting sample. According to Bonifacio et al., we use a laser model to describe equations of motion as

$$\dot{\rho}(t) = -\mathrm{i}(L_{\mathrm{AF}} + i\Lambda_{\mathrm{F}})\rho(t) \,, \tag{138}$$

where L_{AF} denotes the Liouvillian operator, and Λ_{F} is the dissipation term that represents photons coming away from the cavity. They are defined as follows:

$$L_{\mathrm{AF}}\rho = \frac{1}{\hbar}[H_{\mathrm{AF}}, \rho] \,, \tag{139a}$$

$$H_{\mathrm{AF}} = g(aR_{+} + a^{\dagger}R_{-}) \,, \tag{139b}$$

$$\Lambda_{\mathrm{FX}} = \gamma\{[a\rho, a^{\dagger}] + [a, \rho a^{\dagger}]\} \,, \tag{139c}$$

where the constants (g and γ) and the operators have their usual meanings, as before. If we multiply (138) on the left by a, R_{+}, and R_3, respectively, and if we take the trace, we obtain equations for $\alpha(t) = \mathrm{i}\langle a(t)\rangle$, $r_1(t) = \langle R_{+}(t)\rangle$, and $r_3(t) = \langle R_3(t)\rangle$ as

$$\dot{r}_1 = 2g\alpha r_3 \,, \tag{140a}$$

$$\dot{r}_3 = -2g\alpha r_1 \,, \tag{140b}$$

$$\dot{\alpha} = gr_1 - \gamma\alpha \,, \tag{140c}$$

where we have used the semiclassical approximation to neglect quantum correlations. Introducing the polar angle θ of a Bloch vector as

$$r_1(t) = \frac{N}{2}\sin\theta \,, \quad r_3(t) = \frac{N}{2}\cos\theta \,, \tag{141}$$

one finds the following differential equation that shows the damped harmonic oscillation [60, 63]

$$\ddot{\theta}(t) + \gamma\dot{\theta}(t) = Ng^2 \sin\theta(t) \,, \tag{142}$$

and the equations for conservation of the magnitude of the Bloch vector R^2 and energy of the system

$$\partial_t(r_1^2 + r_3^2) = 0 \,, \tag{143a}$$

$$\partial_t(\alpha^2 + r_3) = -2\gamma\alpha^2 \,. \tag{143b}$$

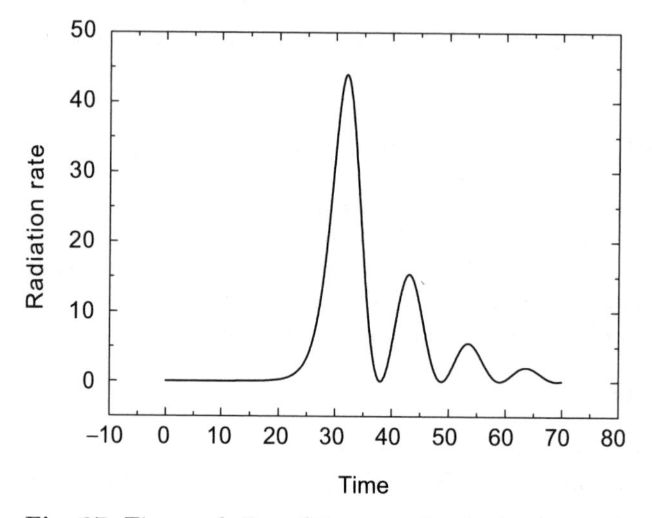

Fig. 27. Time evolution of the rate of radiation from a large sample. The vertical axis is represented in units of the Einstein A coefficient as before. The parameters used are $\gamma = 0.1$, $g = 1/14$, and $N = 20$

It follows from (143b) that the dissipation energy from the system per unit time is $2\gamma\alpha^2$, and it corresponds to radiation energy from the system. Thus we can write radiation intensity $I(t)$ as

$$I(t) = 2\gamma\alpha^2 = \frac{\gamma}{2g^2}\dot{\theta}^2 . \tag{144}$$

In Fig. 27, as an example, we show a numerical result obtained from (142) and (144), where the parameters $\gamma = 0.1$, $N = 20$, and $g = 1/14$ are used. This figure clearly illustrates an oscillation of large-sample superradiance that was first predicted by Bonifacio et al. [60, 63].

If we assume that the reabsorption of photons emitted from the system can be negligibly small enough to drop the $\ddot{\theta}(t)$ term, (142) can be reduced to

$$\dot{\theta}(t) = \frac{Ng^2}{\gamma} \sin\theta(t) . \tag{145}$$

Multiplying $\sin\theta$ on both sides and setting $z = \cos\theta$, we can derive

$$\dot{z} = -\frac{Ng^2}{\gamma}(1 - z^2) , \tag{146}$$

which is the same as (115), the Dicke master equation for a small system obtained semiclassically.

At the end of this subsection, we briefly comment on experimental observation of superradiance. Skribanovitz et al. [64] experimentally showed large sample superradiance similar to Fig. 27, using optically pumped HF atomic gas. For a small system, the radiation profile from CuCl quantum dots was shown to have a single pulse [66].

7 Radiation from the Dipole-Ordered States

In this section we examine the radiation property of the dipole-ordered states discussed in Sects. 4.2 and 5.2. Our model has excluded radiation fields or free photons from the preceding discussions about the dynamics because the interaction energies of excitons with optical near fields are much larger in our system than those with radiation fields. This means that radiation fields are weakly coupled with our system so as not to disturb the dynamics of it. On the basis of such an understanding, we investigate radiation from a quantum-dot system whose dynamics is driven by optical near fields.

7.1 Radiation Property of the Dipole-Ordered States

According to the formulation developed in Sect. 6, we now investigate the radiation property of the dipole-ordered states discussed in Sects. 4.2 and 5.2. From the expression (72) the radiation intensity at time t is determined by the radiation factor, $\langle R_+R_-\rangle = (l+m)(l-m+1)$, i.e., the expectation value of the operator

$$R_+R_- = R_1^2 + R_2^2 + R_3 = \left(\frac{P}{2}\right)^2 + \left(\frac{V}{2}\right)^2 + \frac{W}{2}\,, \tag{147}$$

which indicates that there are two elements mainly contributing to the radiation factor; one is the collectiveness of the system measured by the cooperation number l, and the other is the total energy of the quantum-dot system given by $\langle R_3\rangle = m$. Since the dynamics of the system is driven in our system by localized photons, the evolution of the radiation factor is also described in terms of $l(t)$ and $m(t)$ that are developed according to the localized photon–exciton interaction.

Numerical results of (46) and (47) are shown in Figs. 28 and 29. The upper parts in Fig. 28 show the time evolution of the radiation factors, while the lower parts illustrate the dipole distribution when each radiation factor has a maximal value. The total dipole in Fig. 28a is smaller than those in Figs. 28b and 28c, and the peak value of the radiation factor is also smaller. It follows from the figure that the radiation factor increases as the total dipole becomes larger. The peak values of the radiation factor in Figs. 28b and 28c, 15 and 14, correspond to the value for the Dicke's superradiance, which is obtained as 20 for $N = 8$ from (74). We thus expect that quasisteady states shown in Figs. 28b and 28c are close to the superradiant states, and that the total Bloch vectors for such states are on the $P - V$ plane.

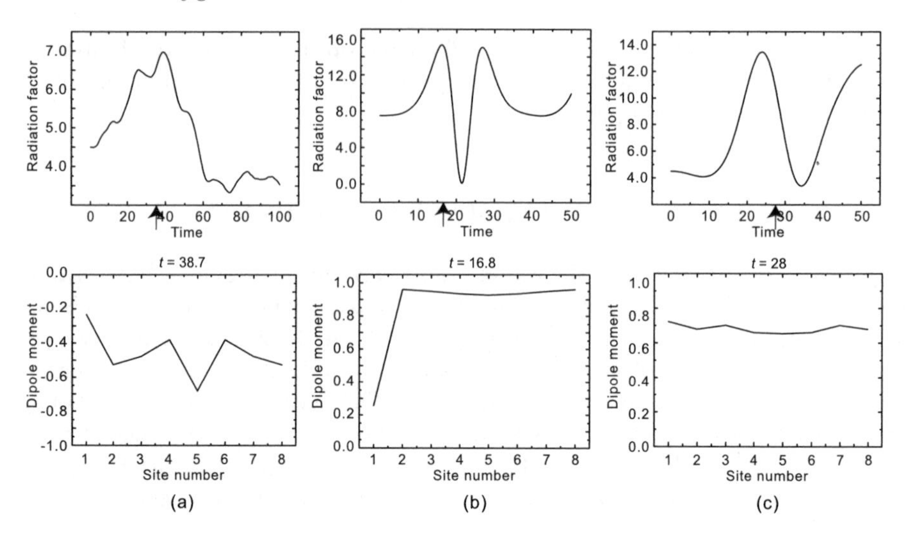

Fig. 28. Time evolution of the radiation factor $\langle R_+R_-\rangle$ (upper) and the dipole distribution at time indicated by the arrows when the radiation factors take maximal values (lower) for (**a**) a system belonging to the first group with an initial population difference $W_n = (0++---++)$, (**b**) a system belonging to the third group with an initial population difference $W_n = (0+++++++)$, and (**c**) a system belonging to the fourth group with an initial population difference $W_n = (0-+-+-+-)$

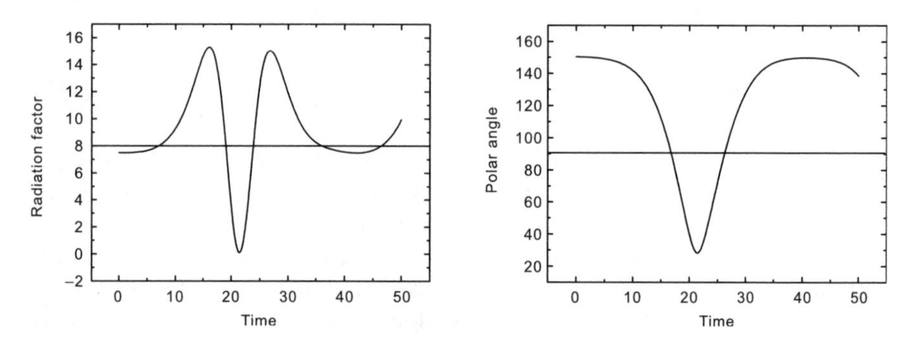

Fig. 29. Time evolution of the radiation factor (*left*) and the polar angle of the total Bloch vector (*right*) for a system belonging to the third group. The polar angle is measured from the $W_n = (--------)$ axis

In order to check whether the total Bloch vectors for the states belonging to the third group in Table 1 (see Fig. 28b) are on the $P-V$ plane or not, we examine the time evolution of the polar angles of the Bloch vectors as well as the radiation factors. As shown in Fig. 29, the polar angle of the Bloch vector takes 90 degrees, that is, on the $P-V$ plane when the radiation factor has a maximal value. Therefore we conclude that the system belonging to the third group in Table 1 is in transition to a quasisteady state close to

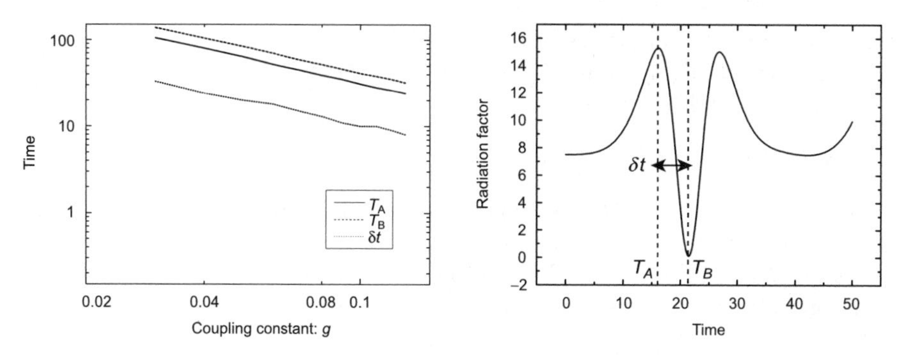

Fig. 30. The dependence of the width and the peak point of superradiant pulse on localized photon–exciton couplings g: double logarithmic plot (*left*) and the definitions of the pulse width δt, the pulse peak points T_{A}, and the bottom point of pulse valley T_{B} (*right*)

the superradiant Dicke state, judging from its large radiation factor and the polar angle of the total Bloch vector.

In order to see the mechanism of the superradiance phenomena in our model we shall investigate the dependence of the localized photon–exciton couplings g on the width of superradiant pulse. The result is shown in Fig. 30.

From this plot we find a relation as follows

$$T \propto \frac{1}{g} \, . \tag{148}$$

Recalling the mechanism of the oscillation of superradiant pulse in the Dicke model explained by Bonifasio (see (142)) we know that this result means the oscillation of superradiant pulses in our model is caused by emission and absorption of localized photons.

Figures 28b and 28c indicate that multiple peaks appear in the radiation, or that multiple pulses are emitted from the system. One may think, as a possible origin, that such a phenomenon stems from the recurrence inherent in an isolated system. However, such multiple pulses may survive even if the system becomes dissipative, which will be examined in detail in the next section.

8 Radiation from a Dissipative System

We have assumed in the previous section that the radiation field is so weak as not to disturb the exciton dynamics of a quantum-dot system. When radiation pulses are emitted from the system, however, energy has to be dissipated, and it is interesting but not clear whether multiple pulses shown in Figs. 28b and 28c are emitted from the system or not. In this section we thus study the radiation profile, adding a radiation field to the system as a reservoir that does not affect the dynamics but makes the system a dissipative one.

Suppose the Hamiltonian

$$H_2 = H_{\mathrm{QDeff}} + H_{\mathrm{F}} + H_{\mathrm{Fint}} , \tag{149}$$

where H_{QDeff} is the effective Hamiltonian given by (51a) that describes the N two-level quantum-dot system interacting with localized photons. The Hamiltonians H_{F} and H_{Fint} describe the free-radiation field and the exciton–free photon interaction, respectively. Explicit Hamiltonians are written as

$$H_{\mathrm{F}} = \sum_{\boldsymbol{k},\lambda} \hbar\omega_{\boldsymbol{k}} a^{\dagger}_{\boldsymbol{k},\lambda} a_{\boldsymbol{k},\lambda} , \tag{150}$$

$$H_{\mathrm{Fint}}(t) = \sum_{\boldsymbol{k},\lambda} \hbar g_{\boldsymbol{k},\lambda} a^{\dagger}_{\boldsymbol{k},\lambda} R_{-} \mathrm{e}^{\mathrm{i}(\omega_{\boldsymbol{k}}-\Omega)t} + h.c. , \tag{151}$$

where creation and annihilation operators of a free photon with wavevector $\boldsymbol{k}$, polarization λ, and frequency $\omega_{\boldsymbol{k}}$ are denoted as $a^{\dagger}_{\boldsymbol{k},\lambda}$ and $a_{\boldsymbol{k},\lambda}$, respectively. The coupling constant between the free photon and exciton is given as

$$g_{\boldsymbol{k},\lambda} = \frac{\mathrm{i}\Omega}{\hbar\sqrt{V}} \sqrt{\frac{\hbar}{2\omega_{\boldsymbol{k}}\epsilon_0}} \boldsymbol{\mu} \cdot \boldsymbol{\epsilon}^{*}_{\boldsymbol{k},\lambda} , \tag{152}$$

where V, $\epsilon^{*}_{\boldsymbol{k},\lambda}$, and ϵ_0 represent the quantization volume of the radiation field, the unit polarization vector, and the dielectric constant in vacuum, respectively. Using the Hamiltonian H_2, we write the Liouville equation as

$$\frac{\partial \rho(t)}{\partial t} = \frac{1}{\mathrm{i}\hbar}[H_2, \rho(t)] = -\mathrm{i}L_2 \rho(t) , \tag{153}$$

and eliminate the degrees of freedom of the radiation field with the help of a projection operator defined as [77]

$$\mathcal{P} \cdots = |0\rangle\langle 0| \mathrm{Tr}_{\mathrm{F}} \cdots , \tag{154}$$

with the vacuum of the radiation field $|0\rangle$. Moving to the interaction representation, we obtain equations of motion for the density operator $\tilde{\rho}(t)$ as follows:

$$\begin{aligned}\mathcal{P}\frac{\partial \tilde{\rho}}{\partial t} &= -\mathrm{i}\mathcal{P}L_2(t)\mathcal{P}\tilde{\rho}(t) - \mathrm{i}\mathcal{P}L_2(t)U(t,0)(1-\mathcal{P})\tilde{\rho}(0) \\ &\quad -\mathcal{P}L_2(t)\int_0^t \mathrm{d}\tau U(t,\tau)(1-\mathcal{P})L_2(\tau)\mathcal{P}\tilde{\rho}(\tau) ,\end{aligned} \tag{155}$$

where $L_2(t)$ is the Liouville operator associated with $H_2(t)$, and the operator $U(t,\tau)$ is defined as

$$U(t,\tau) = \exp\left(-\mathrm{i}(1-\mathcal{P}) \int_{\tau}^{t} L_2(t')\mathrm{d}t' \right) . \tag{156}$$

With the Born–Markov approximation applied to the third term of (155), we obtain the following equation for $\tilde{\rho}_A = \mathcal{P}\tilde{\rho}$ as

$$\frac{\partial \tilde{\rho}_A}{\partial t} = -\mathrm{i}L_{\mathrm{QDeff}}(t)\tilde{\rho}_A(t) - \mathrm{Tr}_{\mathrm{F}} L_{\mathrm{Fint}}(t) \int_0^t \mathrm{d}\tau U_{\mathrm{QDeff}}(t-\tau) L_{\mathrm{Fint}}(t-\tau)|0\rangle\langle 0|\tilde{\rho}_A(t) \,, \tag{157}$$

with

$$U_{\mathrm{QDeff}}(t) = \exp\left(-\mathrm{i}L_{\mathrm{QDeff}}t\right) \,, \tag{158}$$

where L_{QDeff} and L_{Fint} are the Liouville operators associated with the Hamiltonians H_{QDeff} and H_{Fint}, respectively. Moreover, using the Born approximation that neglects the exciton operators of higher than the second order [77], we approximate

$$U_{\mathrm{QDeff}}(t) \sim 1 \tag{159}$$

to obtain a compact equation as

$$\frac{\partial \tilde{\rho}_A}{\partial t} = -\mathrm{i}L_{\mathrm{QDeff}}(t)\tilde{\rho}_A(t) + \beta\left\{[R_-\tilde{\rho}_A(t), R_+] + [R_-, \tilde{\rho}_A(t)R_+]\right\} - \mathrm{i}\gamma[R_+R_-, \tilde{\rho}_A(t)] \,, \tag{160}$$

with

$$\beta + \mathrm{i}\gamma \equiv \int_0^t \sum_{\boldsymbol{k},\lambda} |g_{\boldsymbol{k},\lambda}|^2 \mathrm{e}^{\mathrm{i}(\Omega - \omega_{\boldsymbol{k}})\tau} \mathrm{d}\tau \,, \tag{161}$$

which is exactly the same form as Lindblad's master equation [86] describing a general Markovian dynamics for a dissipative quantum system. Here the real and imaginary parts of the right-hand side of (161) are designated as β and γ, respectively. In the following we neglect the energy shift as $\gamma = 0$, for simplicity. Note that the second and third terms of the right-hand side of (160) are known as Dicke's master equation [57,60,62].

8.1 Semiclassical Description with the Effective Hamiltonian

Neglecting quantum correlation between excitons [58], we approximate the total density operator $\tilde{\rho}_A$ as a direct product of the density operator ρ_n at each site n

$$\tilde{\rho}_{\mathrm{A}} = \prod_n \rho_n \,, \tag{162}$$

and then solve (160). Noticing that the dynamics governed by the original Hamiltonian (4) is not rigorously identical to the one described by the effective Hamiltonian (51a), we use an isolated system described by the effective

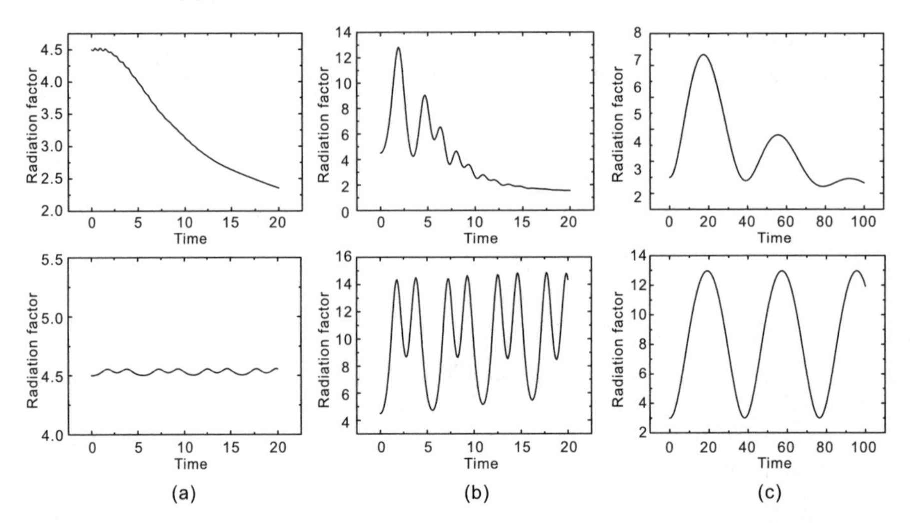

Fig. 31. Time evolution of the radiation factor for dissipative systems (*upper*) and isolated systems (*lower*). We consider the following three cases for both systems: (**a**) a "ferromagnetic" case, (**b**) an "antiferromagnetic" case that is turned into a dipole-ordered state after manipulating the initial population differences, and (**c**) a "dipole-forbidden" case discussed in Sect. 4.3. The parameters $\beta = 0.05$ and $\gamma = 0$ are used

Hamiltonian (51a) in order to clarify the dynamics governed by (160) for a dissipative system whose relevant system is described by the same Hamiltonian (51a). By comparing the radiation factors for the isolated system and the dissipative system, the similarity and the difference are discussed.

Figure 31 shows the time evolution of the radiation factor for a "ferromagnetic" system belonging to the third group of Table 1, an "antiferromagnetic" system that belongs to the fourth group of Table 1 and is turned to a dipole-ordered state after manipulating the initial population differences, and a "dipole-forbidden" system that is converted to a dipole-ordered state by the local manipulation discussed in Sect. 4.3 as $\langle W_n(0)\rangle = -\langle P_n(0)\rangle$. It follows from the lower part of Fig. 31a that superradiance is suppressed and does not manifest itself due to the difference between the original Hamiltonian (4) and the effective Hamiltonian (51a). On the other hand, superradiant multiple pulses are generated in the isolated system and survive in the dissipative system for both cases shown in Figs. 31b and 31c. Therefore we find with a semiclassical approach that multiple pulses, as predicted for an isolated system, can be emitted superradiantly from dipole-ordered states even in a dissipative system coupled to a radiation reservoir.

8.2 Quantum Correlations

It is well known that superradiance in the Dicke model [49] occurs from a state where all excited states of all sites are occupied. On the other hand, the semiclassical approach discussed above cannot predict the occurrence of superradiance of the system when the total dipole of the system is zero as an initial condition [58]. This means that quantum fluctuations and correlations should be properly included so as to correctly describe the radiation properties of a system with no initial dipoles, and that the semiclassical approximation is not appropriate in this case. Thus we numerically solve the master equation (160), taking quantum correlations into account, and we investigate the radiation properties of the dissipative system. In particular, we compare the results obtained from our model, i.e., the localized photon model with those obtained from the Dicke model, for which the first term of the right-hand side is dropped from (160). Some of such results are shown in Figs. 32 and 33.

Figure 32a shows the time evolution of the radiation factor for the case that all the populations are completely in the excited states and there are no dipoles as an initial condition. The solid curves are the results for our model, while the dashed curves represent the results for the Dicke model. It follows from the figure that a single superradiant pulse is emitted in both models, but that the peak value of the superradiant pulse is reduced, while the tail is extended in our model. In Fig. 32b, we present the result obtained from the initial condition as a zigzag profile of the dipole distribution of $\langle P_n \rangle = (-1)^n/\sqrt{2}$ and flat population differences of $\langle W_n \rangle = 1/\sqrt{2}$. The system corresponds to a dipole-forbidden state as shown in Fig. 11, where the total dipole is always zero. The radiation profiles obtained for both models are qualitatively, same as found as in Fig. 32a. The time evolution of the radiation factor is illustrated in Fig. 32c for the case that initially the alternating dipole is set as $\langle P_n \rangle = (-1)^n/\sqrt{2}$ and the sign of the population difference in each QD is set opposite to that of the corresponding dipole as $\langle W_n \rangle = -\langle P_n \rangle$. Owing to the flip hypothesis, a dipole-ordered state emerges in this case, and a distinct difference between the two models is observed: our model (solid curve) shows superradiance, while the Dicke model (dashed curve) does not. Since we can infer that the difference stems from the dipole-ordering phenomenon, we further examine this case in order to clarify the relation between the radiation factor and the total dipole of the system.

Figure 33 presents the time evolution of the radiation factor (upper) and the total dipole (lower) by changing the exciton–localized-photon coupling constant g as $g = 0.5$, 0.8, and 1.2. It is found that the frequency of the total dipole increases as the coupling constant becomes strong. As a result, the oscillation frequency of the radiation factor increases, and thus multiple pulses are emitted superradiantly from the dissipative system. Therefore the difference between the two models, as we expected, originates from the occur-

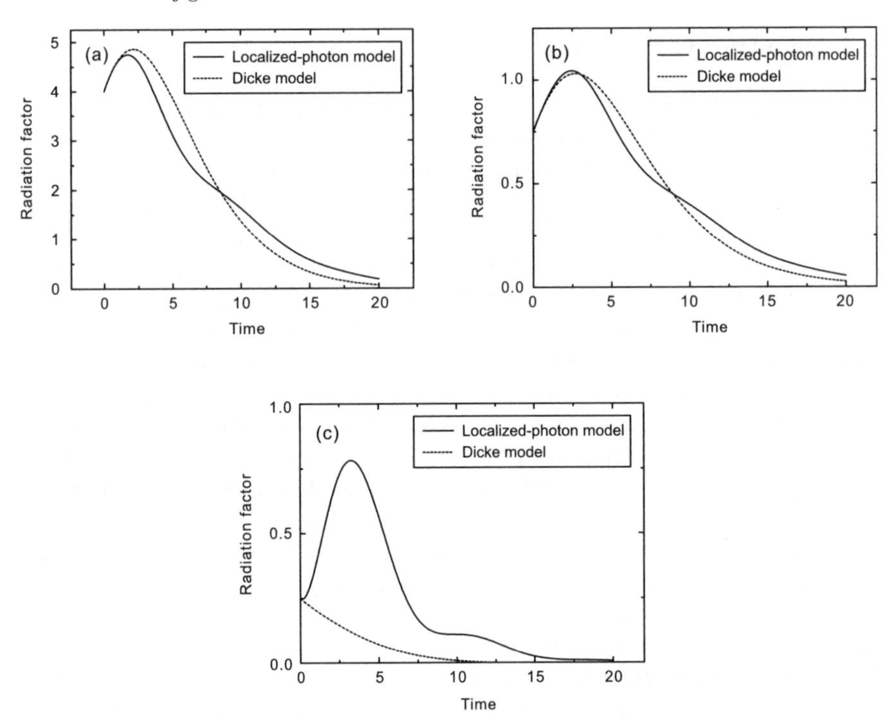

Fig. 32. Time evolution of the radiation factor obtained with quantum correlations. The *solid curves* are the results for our model, while the *dashed curves* represent the results for the Dicke model. The dissipative system is assumed, and the parameters $N = 4$, $\beta = 0.05$, and $\gamma = 0$ are used. In addition, the following initial conditions are used: (**a**) $\langle P_n \rangle = 1$ and $\langle W_n \rangle = 1$, (**b**) $\langle P_n \rangle = (-1)^n/\sqrt{2}$ and $\langle W_n \rangle = 1/\sqrt{2}$ which correspond to a dipole-forbidden state as shown in Fig. 11, and (**c**) $\langle P_n \rangle = (-1)^n/\sqrt{2} = -\langle W_n \rangle$, which corresponds to a dipole-ordered state as shown in Fig. 12

rence of a dipole-ordered state or a collective dipole oscillation via localized photon-exciton interactions.

At the end of this section, we examine the applicability of the semiclassical approach that has an advantage over the quantum approach that one can easily handle a relatively large number N system. In Fig. 34 we show the radiation profile obtained from three different methods. The system and the initial conditions to be considered are the same as in Fig. 33. The result is obtained for the isolated system by the Liouvillian dynamics of the effective Hamiltonian with the semiclassical approximation, for the dissipative system by the master equation (160) with the semiclassical approximation, and for the dissipative system by the master equation (160) with quantum correlations. We find from Fig. 34 that multiple pulses are generated for all the cases though each frequency of the oscillation is different. This indicates

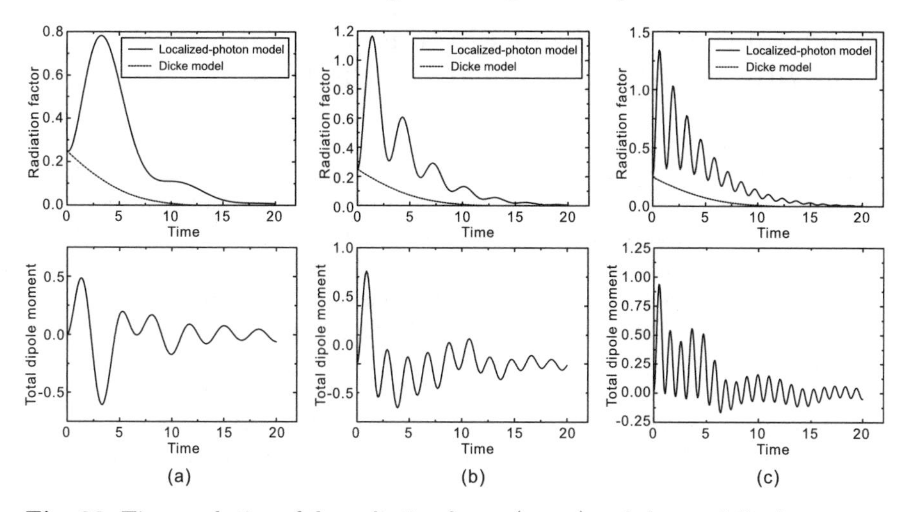

Fig. 33. Time evolution of the radiation factor (*upper*) and the total dipole moment (*lower*) obtained with quantum correlations. The *oscillating curves* are the results for our model, while the *monotonically decreasing curves* represent the results for the Dicke model. The system is assumed to be dissipative and in a dipole-ordered state evolved from a dipole-forbidden state, i.e., the same as in Fig. 32c. The initial conditions are also the same as in Fig. 32c, except for the exciton–localized-photon coupling (**a**) $g = 0.5$, (**b**) $g = 0.8$, and (**c**) $g = 1.2$

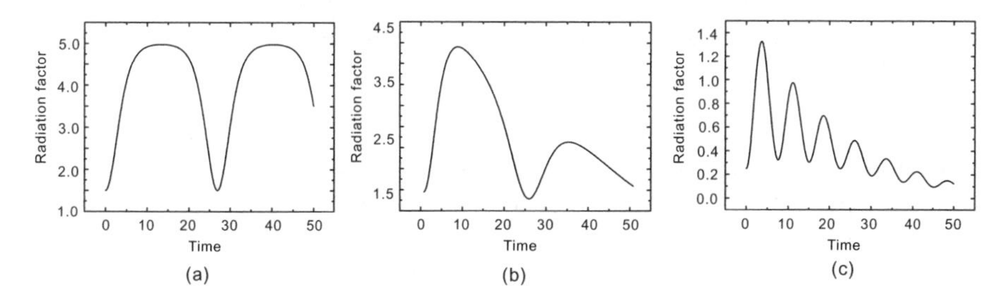

Fig. 34. Radiation profile obtained from three different methods. The system and the initial conditions to be considered are the same as in Fig. 33. The result is obtained (**a**) for the isolated system by the Liouvillian dynamics of the effective Hamiltonian with the semiclassical approximation, (**b**) for the dissipative system by the master equation (160) with the semiclassical approximation, and (**c**) for the dissipative system by the master equation (160) with quantum correlations. The parameters $N = 4$, $\beta = 0.05$, and $\gamma = 0$ are used

that the semiclassical approach can describe qualitatively the radiation properties of both isolated and dissipative systems when the total dipole of the system is not zero. The strong radiation coming from the dipole-ordering phenomenon, or the nonlinearity and the collective phenomena of the dynamics of the system considered in this chapter can be qualitatively predicted by the semiclassical approach.

9 Dynamics of Localized Photons: the Storage Mode of Localized Photons

Since the semiclassical Heisenberg equations (46) are nonlinear, neglecting quantum correlations between quantum dots (QDs) and localized photons, we expect such a system to display nonlinear or possibly chaotic phenomena. Indeed, since a system with strong nonlinearity behaves chaotically and shows deterministic randomness, chaotic systems represent macroscopic transport phenomena. In this context, this section considers an open system for the localized photon model and investigates the transportation of localized photons.

As depicted in Fig. 35, we lined up QDs that interact with localized photons with a reservoir of localized photons at either end of the line. In this setting, we investigated the transportation of localized photons through the system.

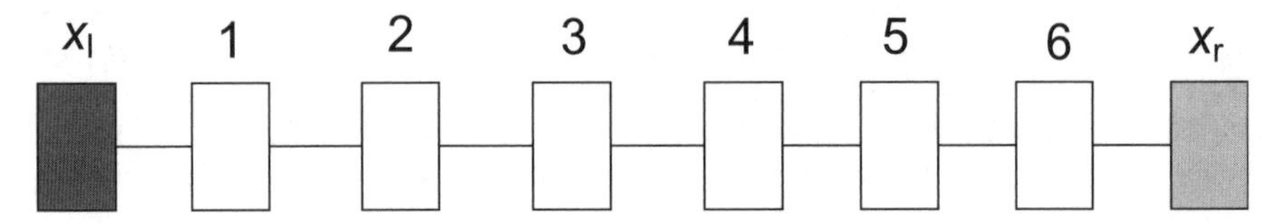

Fig. 35. The QDs interacting with localized photons are configured in a line. At each end of the line, we set different boundary conditions

9.1 The Transportation of Localized Photons: the Storage and Through Flowing Modes

The numerical operator of the localized photons at site j is defined as $n_j = a_j^\dagger a_j$ and the equation of motion for the expectation value $\langle n_j \rangle$ using the semiclassical approximation is derived as follows.

$$\frac{\mathrm{d}\langle n_j \rangle}{\mathrm{d}t} = \frac{v}{2}(\langle x_{j-1} \rangle + \langle x_{j+1} \rangle)\langle y_j \rangle \\ -\frac{v}{2}(\langle y_{j-1} \rangle + \langle y_{j+1} \rangle)\langle x_j \rangle - \frac{g}{2}(\langle V_j \rangle - \langle P_j \rangle)\langle y_j \rangle \,. \quad (163)$$

Since the right-hand side of (163) does not include the localized photon number $\langle n_j \rangle$, the dynamics of the localized photon number are decided by the other Heisenberg equations (46) and (47). From the Heisenberg equations, it is clear that using the semiclassical approximation, the physical quantity that can be input from both ends as a boundary condition is the coherence of localized photons $x_j = a_j + a_j^\dagger$ and $y_j = i(a_j - a_j^\dagger)$ (not the localized photon number $a_j^\dagger a_j$). Therefore, we define the boundary condition at both ends using (57) to achieve a coherent state in which the eigenvalue is a real number.

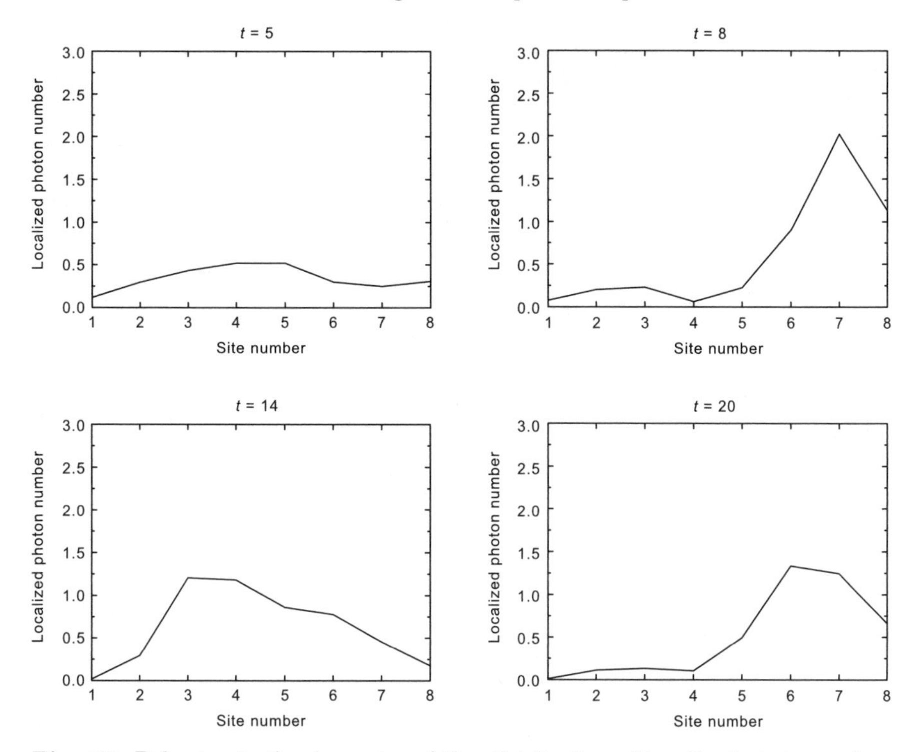

Fig. 36. Behavior 1: the dynamics of the distribution of localized photons of an open system with reservoirs of localized photons in the coherent states at the zeroth and ninth sites. In this case we observe linear propagation and the reflection of excitation of localized photons

This boundary condition corresponds to the locally determined amplitude of the classical electric field oscillation (at the boundaries). The electric field at the left boundary is larger than that at the right boundary ($x_l > x_r$). The emerging dynamics are then examined with the focus on the distribution of localized photons. Consider the case where there are eight sites and the values at sites 0 and 9 are fixed at $\langle x_0 \rangle = 1$, $\langle x_9 \rangle = 0$, respectively, by assuming that a reservoir of localized photons has been attached, with all the excitons in the initial state of excitation, and localized photons are set to 0 at all the other locations, except at the boundaries. In this configuration, the temporal evolution of localized photons shows that the resultant behavior can be broadly classified into the following two types: Behavior 1 is shown in Fig. 36, in which the excitation travels from left to right, just like a wave. Behavior 2 is shown in Fig. 37, in which a structure similar to a standing wave emerges and develops over time.

In Fig. 36, transmission of a wave can be observed in which excitation emerges from the input side, travels to the right, reflects off the right boundary, travels to the left, gets pushed back by the input on the left, and travels

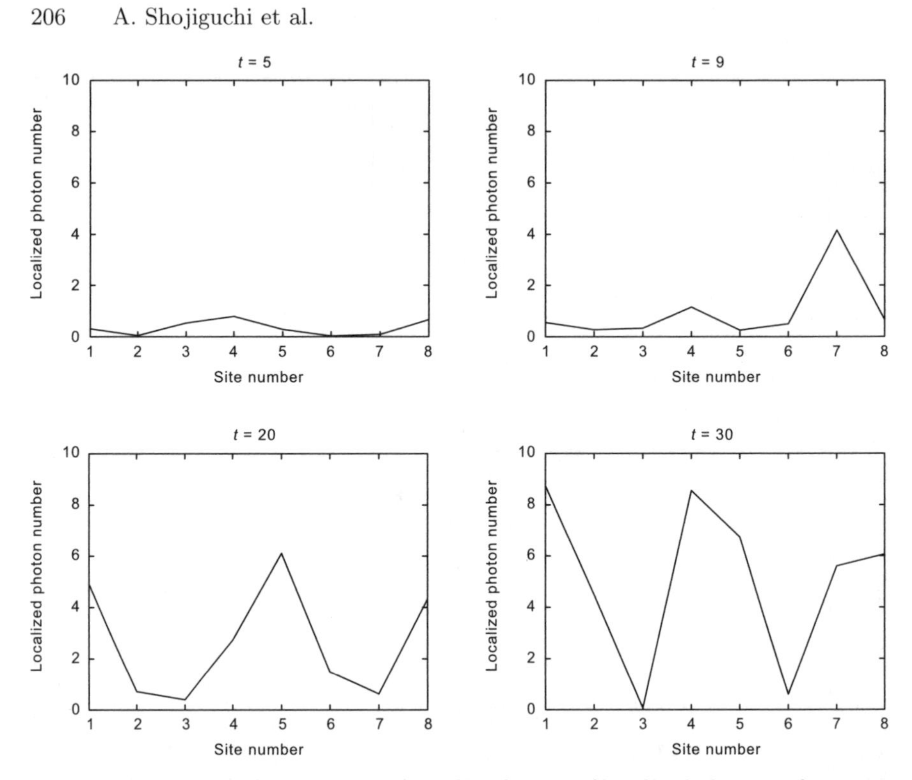

Fig. 37. Behavior 2: the dynamics of the distribution of localized photons of an open system with reservoirs of localized photons in the coherent states at the zeroth and ninth site. In this case, we observe the growth of a structure similar to a standing wave

to the right again. Consequently, it travels back and forth. In other words, linear behaviors like transmission of a wave can be observed in the temporal evolution of localized photons despite the nonlinear nature of the system's dynamics. In Fig. 37, there is a clear division between the sites where localized photons accumulate (peaks) and the sites where localized photons are rare (nodes). Localized photons continue to accumulate in the system and no stationary state emerges. This can be viewed as formation of a standing wave.

Examination of the total number of localized photons $\sum_n \langle a_n(t)^\dagger a_n(t) \rangle$ in the system yields the results shown in Fig. 38.

Figure 38 clearly distinguishes between the two types of system when looking at the number of photons. In the system shown in Fig. 36, in which the transportation of localized photons follows a linear wave motion, the total number of photons is approximately constant. This means that localized photons in the system are flowing through the system. Therefore we call this through flowing mode. In the system shown in Fig. 37, the total number of localized photons increases monotonically and localized photons continue to accumulate in the system. We call this phenomenon the photon storage

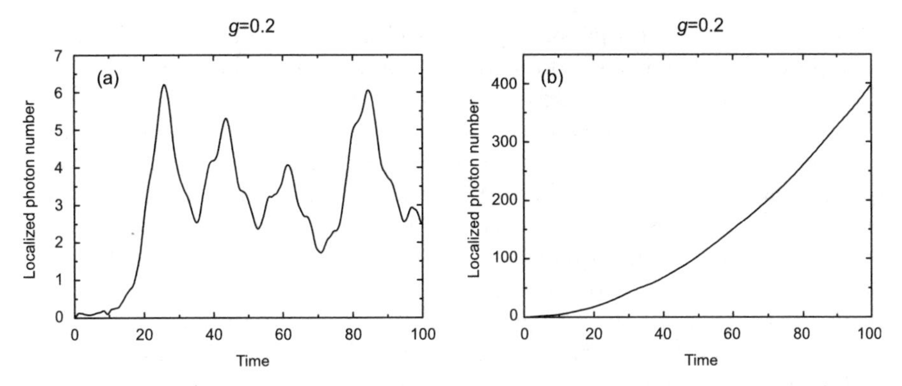

Fig. 38. Temporal evolution of the total localized photon number $\sum_n \langle a_n(t)^\dagger a_n(t)$ (feet). (**a**) and (**b**) correspond to the systems in Fig. 36 and Fig. 37, respectively

mode. The fact that the relaxation phenomenon does not occur implies that either photons can accumulate in the system, or that the system relaxes very slowly and our calculations are insufficient for it to reach equilibrium.

Following the discussion given in Sect. 7 and using (147), we then calculated the radiation intensity for the two systems, as shown in Fig. 39. It follows from Fig. 39 that no cooperative effect can be seen in the system shown in Fig. 36, while a cooperative effect (superradiance) and a time-irreversible phenomenon (dacaying profile of radiation) is observed in the system shown in Fig. 37. Here the system parameters used in Fig. 37 are same as those used in Fig. 28b, but it should be noted that a superradiant radiation intensity decays time-irreversibly in the open system as assumed in Fig. 37 while a recursive superradiance occurs in the closed system consid-

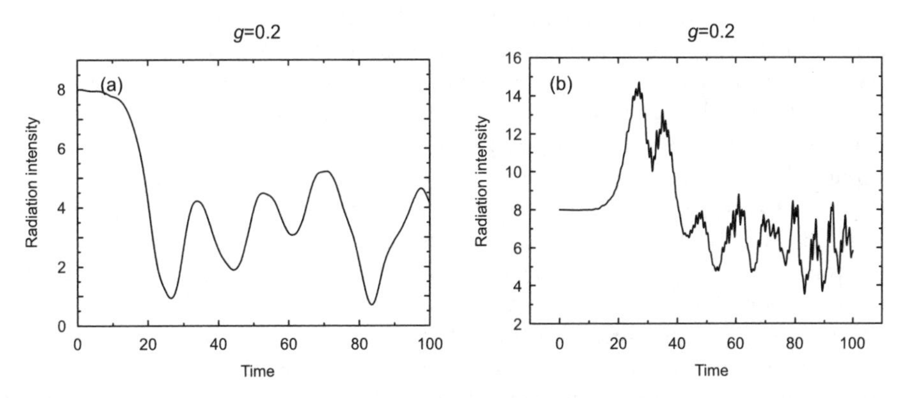

Fig. 39. Temporal evolution of the radiation intensity of the open systems. (**a**) and (**b**) correspond to the systems in Fig. 36 and Fig. 37, respectively. Strong radiation occurs in the system that is accumulating the localized photons and not in the system where there is no accumulation

ered in Fig. 28b. Nevertheless, it becomes clear that a statistical-mechanical transportation phenomenon does not occur as a nonequilibrium steady state, despite the nonlinear dynamics of the system. One of the causes of this is that the system's nonlinearity is weak. Generally, the ergodicity or mixing property strengthens as nonlinearity increases, and relaxation is reached more quickly towards the steady state [87]. Another cause may be that chaotic characteristics in the classical sense do not emerge, because the system is currently dealt with in terms of semiclassical quantum mechanics. However, other phenomena characteristic of a nonlinear system should appear when nonlinearity exists in the system of equations, similar to quantum chaotic systems in which characteristic phenomena different from those of a regular system appear indirectly in the level statistics [88,89]. The following subsection contains observations regarding this point.

9.2 The Emergence of Dynamical Nonlinearity

Dependence of the Coupling Constant g

To study the appearance of nonlinearity in this system, the dependency of the temporal evolution of the localized photons on the coupling constant g is examined. The temporal evolution of the total number of localized photons that accumulate in the entire system for the case in Fig. 38b is plotted for various coupling constants, g, as shown in Fig. 40.

Looking at the plot, there is a characteristic sudden shift in behavior from monotonically increasing to oscillatory at $g = 0.67$, and this returns to monotonically increasing behavior at $g = 0.69$. Indeed, the number of localized photons decreases by one order of magnitude at the transition from monotonic behavior at $g = 0.65$ to oscillatory behavior at $g = 0.67$. This is a kind of nonlinear behavior. That is to say, we can expect nonlinear, random behavior in the dependency of the total number of localized photons on the exciton–photon coupling constant g. To examine this in detail, Fig. 41 plots the dependency of the total number of localized photons on the coupling constant g at time $t = 100$. Figure 41 shows that the appearance of nonlinear, random behavior for the dependency of the total number of localized photons depends on the exciton–photon coupling constant g, for the systems in both Figs. 36 and 37. Particularly for the system in Fig. 37, in which localized photons accumulate, the fluctuation in the dependency of the total number of photons on the coupling constant is extremely vigorous. Examination of the Heisenberg equations (46) and (47) shows that the coupling constant g is included in the nonlinear terms of the equations. Therefore, it is easy to see that the coupling constant g is one of the parameters that measures nonlinearity. Therefore, the dynamics of systems with boson-approximated excitons, in which nonlinear terms do not emerge in the equations of motion, must be examined in order to confirm that this behavior originates from the nonlinearity of the equations.

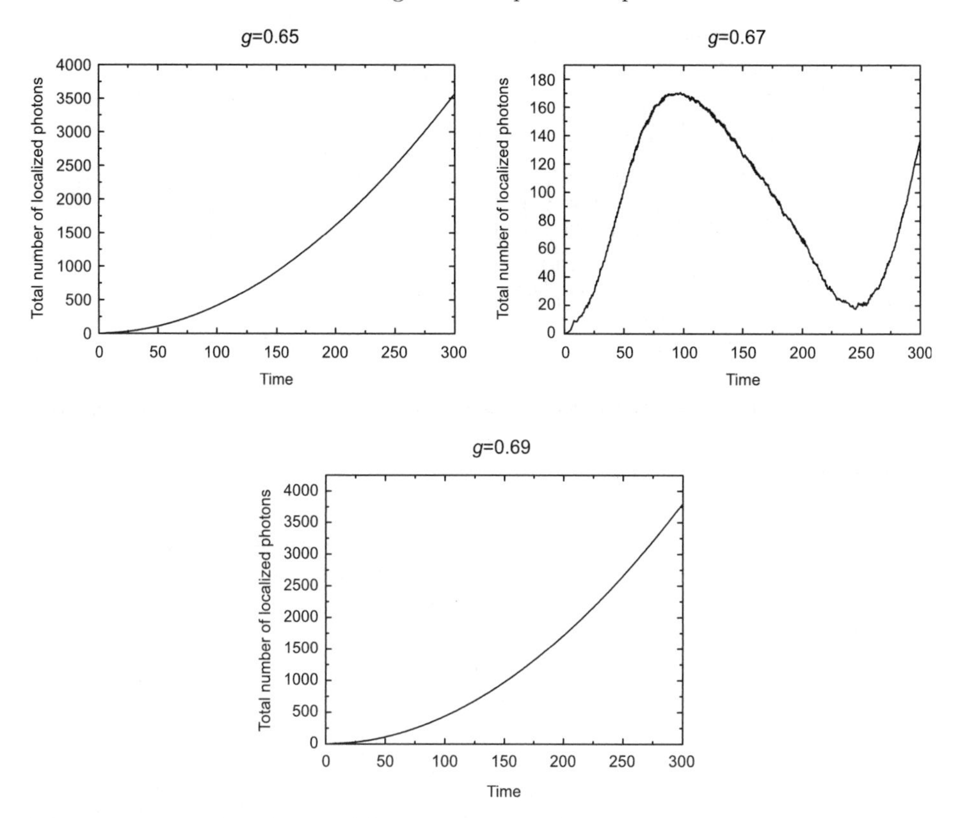

Fig. 40. Temporal evolution of the total number of localized photons in the system corresponding to Fig. 38b when the coupling constant g equals 0.65, 0.67, and 0.69. We observe a drastic change of the number of localized photons between $g = 0.65$ and $g = 0.67$. The difference amounts to a single digit

Linear System

The Heisenberg equations of the system become the following when an exciton is approximated by a boson:

$$\begin{cases} \partial_t P_n = -\Omega V_n - g y_n \,, \\ \partial_t V_n = \Omega P_n + g x_n, \\ \partial_t x_n = -\omega y_n - v(y_{n-1} + y_{n+1}) - g V_n \,, \\ \partial_t y_n = \omega x_n + v(x_{n-1} + x_{n+1}) + g P_n \,. \end{cases} \tag{164}$$

These are clearly linear equations. Since the equation of motion of the number of localized photons $\langle a_n(t)^\dagger a_n(t)\rangle$ is the same as in the case of (163) discussed above, which is obviously linear, in the boson-approximated system

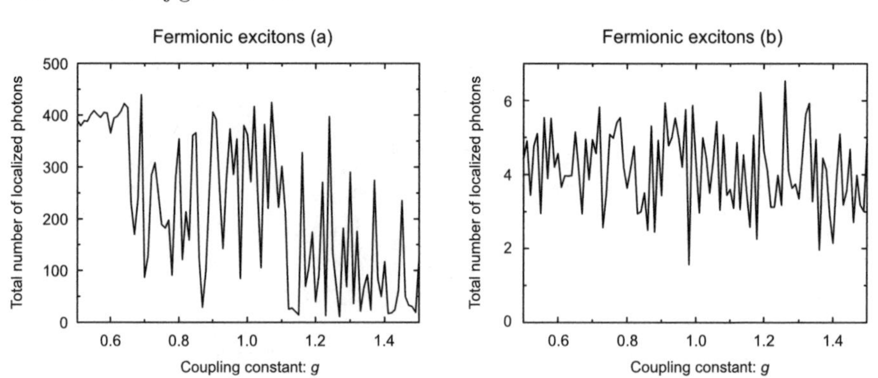

Fig. 41. Dependence of the total localized photon number on the coupling constant g at $t = 100$. (**a**) and (**b**) correspond to the systems in Fig. 36 and Fig. 37, respectively

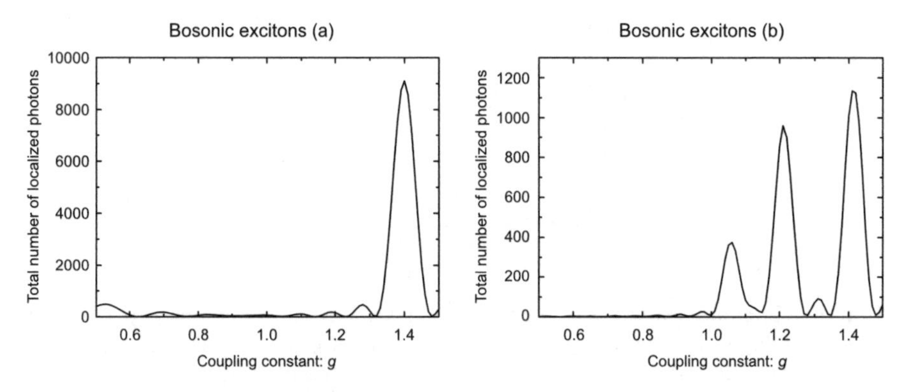

Fig. 42. The dependence of the coupling constant g of the total localized photon number at $t = 100$ for the system in which there is boson-approximation of excitons. (**a**) and (**b**) take the same system parameters as the systems in Fig. 36 and Fig. 37, respectively

all the equations of motion are linear. Solving these equations, Fig. 42 shows the dependency of the total number of localized photons on the coupling constant g. Although random behavior does not appear in this case, resonating coupling constant dependence is found. The meaning of the resonating coupling constants is not yet clear; however, at least there is no disorderly, unpredictable behavior. From these observations, we conclude that the disorderliness of Fig. 41 is a chaotic behavior originating from the nonlinearity of the dynamics.

Here, one must be cautious, because a system consisting of eight sites has many modes, and it is possible that an apparently disorderly dependence has emerged via the competition of many modes (for example, the superposition of many types of quasiperiodic behavior). Quasiperiodic behavior does not have a period either, so chaos does not exist simply because of a lack of periodicity. In the next section, the phenomenon is examined by simplifying the problem by reducing the degrees of freedom to the minimal value, while considering the above.

9.3 Two-Site Open System with Intermittent Chaotic Behavior

This section considers the nonlinear dependency on the coupling constant g observed in the previous section in a system with two sites (four sites including the reservoirs at the two ends) in order to simplify the problem. This is done to reduce the degrees of freedom so as to differentiate between the complexities caused by the superposition of many quasi-periodic behaviors originating from the many modes and the nonlinearity of the dynamics of the system.

Figure 43 shows the dependency of the total number of localized photons on the coupling constant g for the two-site system. This shows that dependency on g behaves unpredictably. Since with two sites there are few participating modes, the vigorous fluctuations are a manifestation of chaotic behavior, reflecting the nonlinearity of the dynamic equations. Examination of Fig. 43a shows that there are vigorous fluctuations (chaotic regions) in some areas, and not in others (regular windows). In other words, this system displays 'intermittent chaos' [87], using the terminology to describe the temporal evolution of chaos. Since chaotic systems display deterministic probabilistic behavior (e.g., diffusion) [87], the number of localized photons should obey a statistical distribution. Therefore, we examined histograms of the number of localized photons corresponding to Figs. 43a–d. Figure 44 shows the histograms of the numbers of localized photons in the chaotic regions of the system corresponding to Figs. 43a–d in a double-logarithmic graph. This figure shows that the histograms for each individual region share a common behavior. The double logarithmic plot of the histograms consists of two straight lines; the first line has a slope of approximately -1. In other words, when the distribution of localized photons n is approximated as $P(n)$, the distribution in the regions of small n is given by $P(n) \propto 1/n$. The slope is very steep in the region where there are many photons, and the graph can be regarded as an exponential function. In conclusion, the emergence of a common statistical distribution of localized photons in regions (**a**) through (**d**) is the evident result of a universal statistical law caused by the chaotic dynamics of the system. This means that the apparent complexity does not arise merely by chance; there is a common principle working in the background. In other words, the nonlinear dynamics of the system indirectly cause such behavior to emerge.

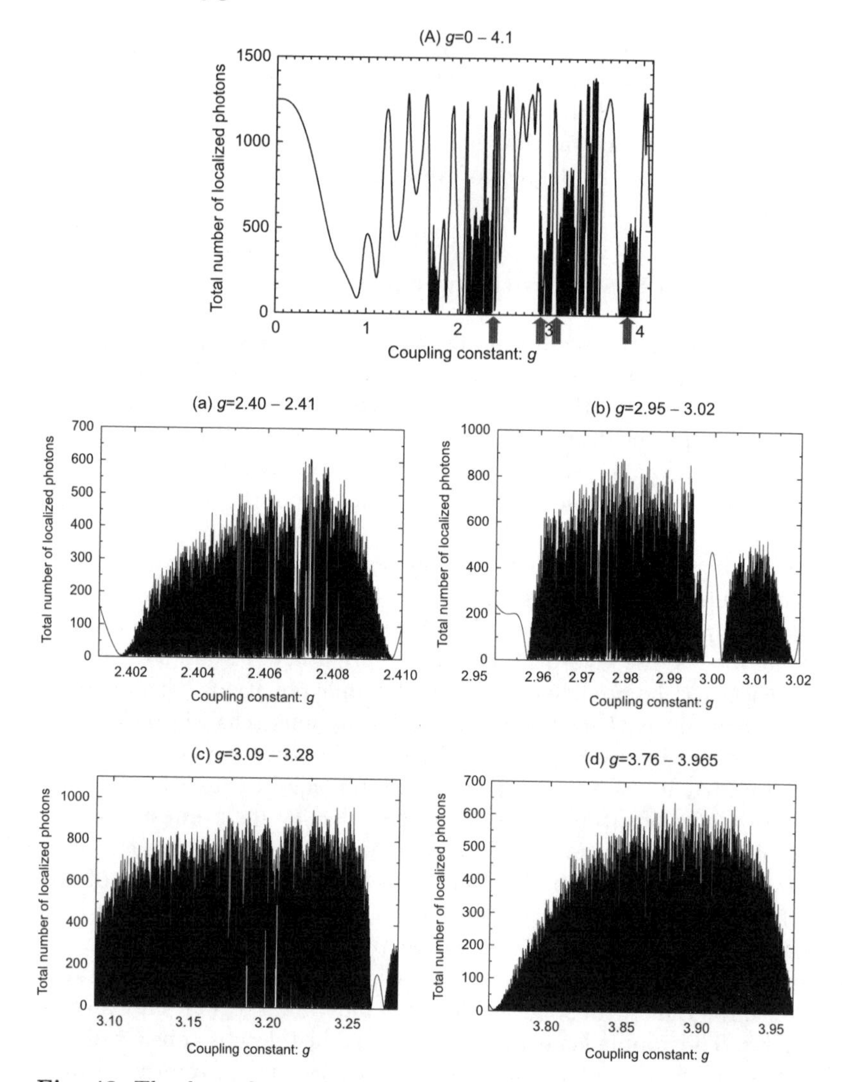

Fig. 43. The dependence of the coupling constant g on the total localized photon-number of the two-site open system. (**A**) is the plot for the range from $g = 0$ to 4.1. (**a**)–(**d**) show the chaotic regions in (**A**). (**a**) shows the range from $g = 2.40$ to 2.41, (**b**) is from $g = 2.95$ to 3.02, (**c**) is from $g = 3.09$ to 3.28, and (**d**) is from $g = 3.76$ to 3.965

As we show above, many interesting phenomena emerge when we look at the behavior of localized photons, including the existence of the photon-storage mode and the existence of chaotic dependency of the system's mode on a system parameter. Nevertheless, our discussion is based on a semiclassical system of equations that ignores quantum correlations, and it remains

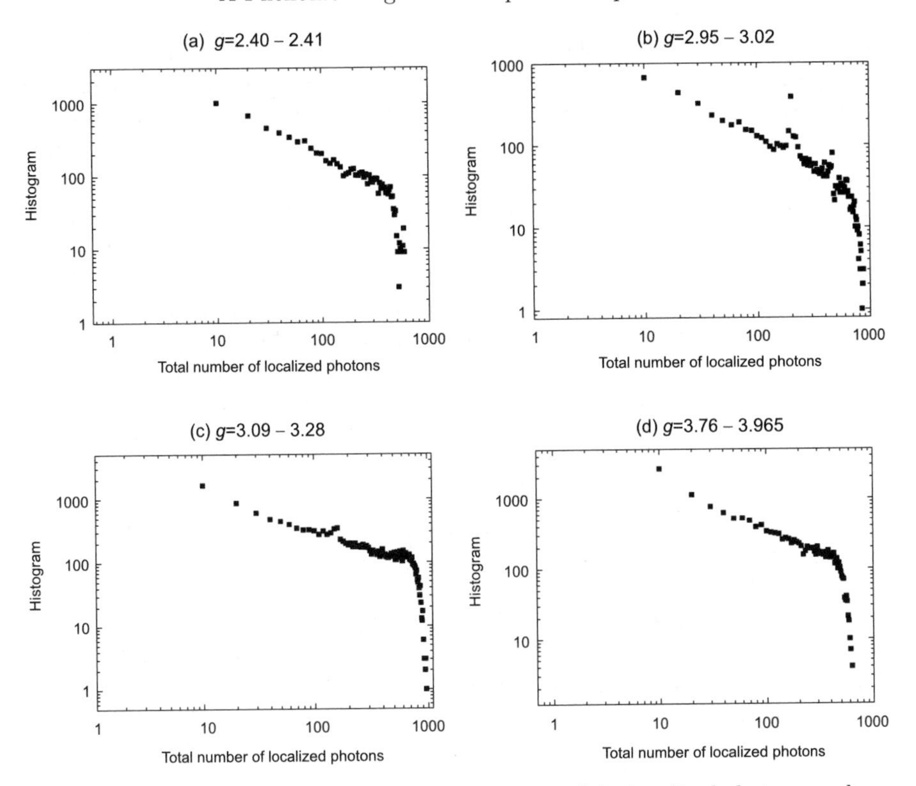

Fig. 44. A double logarithmic plot of the histogram of the localized photon numbers for the regions depicted in Fig. 43. Figures (**a**)–(**d**) correspond to chaotic regions (**a**) to (**d**) in Fig. 43

to be determined whether these phenomena actually occur, i.e., do these phenomena survive in the case using full quantum correlations? Therefore, we examined the behavior considering the quantum correlation of the next order. In this case, the storage mode for photons also exists. The fact that the behavior in the next order does not differ significantly suggests that the semiclassical approach is sufficient.

10 Conclusions

In order to investigate the dynamics of the system, we presented a model of N two-level quantum-dot systems interacting with optical near fields that are explicitly expressed in terms of the localized-photonic degrees of freedom, which characterize the unique property of the localization of optical near fields. At the low-density limit, excitons (N two-level systems) are approximated as bosons, and a rigorous solution of the Heisenberg equation is obtained. Using this solution, we examined the dynamics of the excitonic system, and showed that the dipole moments are transferred linearly within the system. Since the dipole moments in the system represent the quantum coherence between any two energy levels, this phenomenon might be applicable to a photonic device on a nanometer scale or to the transfer of quantum information.

For fermionic excitons, the Heisenberg equation becomes nonlinear, and the dynamics is more complicated. We obtained a perturbative solution given by (43) within the second order with respect to the localized photon–exciton interaction, to investigate the dipole dynamics. The study revealed that several oscillating quasisteady states exist, depending on the material parameters. Using the effective Hamiltonian obtained from renormalization of the localized photonic degrees of freedom, we classified such quasisteady states into several groups: one is a "ferromagnetic" state with all the dipoles aligned in the same direction, the other is an "antiferromagnetic" state with alternating dipoles, as shown in Table 1.

Note that if the sign of the population difference $\langle W_n(0)\rangle$ at an arbitrary site n and time $t = 0$ is reversed, then the dipole of the same site at arbitrary time t, $\langle P_n(t)\rangle$, also changes sign (see (44)). Using this flip hypothesis, we can transform an arbitrary dipole distribution of the system into a dipole-ordered state after manipulating the initial distribution of the population differences. This hypothesis is based on the perturbative solution (44), which determines the sign of $\langle P_n(t)\rangle$ according to the sign of the product of $\langle P_1(0)\rangle\langle W_n(0)\rangle$. It also originates from the fermionic property of excitons, which gives the Heisenberg equations of motion for b_n ($\propto P_n$) as

$$[H_{\mathrm{b-b}}, b_n] = \sum_{m \neq n} \Delta\Omega_{nm} W_n b_m \, . \tag{165}$$

The right-hand side of (165) is proportional to $\langle P_m\rangle\langle W_n\rangle$ for fermions, while it is proportional to $\langle P_m\rangle$ for bosons. Therefore, the occurrence of this kind of nonlinearity for fermionic excitons is a possible origin of the flip hypothesis. Moreover, it is not trivial; this property can be applicable to results obtained from the semiclassical Heisenberg equations (46) and (47).

We examined the radiation property of our system, solving the Heisenberg equations (46) and (47) using the semiclassical approximation. In the investigation, we treated the system as isolated, since the coupling between radiation fields and the sytem is very weak. The numerical analysis showed that dipole-ordered states, which have large total dipole moments, show a large radiation probability comparable to Dicke's superradiance. In particular, it predicted that multiple pulses are emitted from the system superradiantly. In order to verify whether such a phenomenon is inherent in an isolated system, we solved the master equation (160) for a dissipative system using the semiclassical approximation, and found that such multiple pulses in the radiation profile can even survive in a dissipative system coupled to a radiation reservoir.

Multiple superradiant peaks have been observed experimentally in an atomic-gas system [64], and are reported to originate from the dipole–dipole interaction between a two-level system within the framework of the Dicke model [65]. In our model, excitons are coupled with one another indirectly through the interactions with localized photons; therefore, the pulse oscillation in the radiation profile occurs in a mechanism similar to that in the Dicke model.

We solved the master equation (160), considering quantum correlations, and compared the results with those obtained from the Dicke model. When all the populations are initially in excited states, both models produce similar radiation profiles. They differ qualitatively in that the peak value of the radiation pulse in our model is reduced and the tail is extended. This tendency was examined by comparing atomic and Frenkel exciton systems, in which excitons can hop via the dipole–dipole interaction, and the same qualitative difference was reported [69]. Regarding the multiple-pulse generation that we obtained, the Frenkel exciton model also predicts the possibility of the superradiance profile oscillating if the system is initially prepared with a partial population inversion [83]. From our model, we concluded that the superradiant peaks of multiple pulses correspond to the occurrence of a collective dipole oscillation, or a dipole-ordered state.

It is still not clear why the dipole distribution in our system has several quasisteady states, which was predicted in our model using the perturbative expressions, numerical solutions of the Heisenberg equations, and the effective Hamiltonian. Since our system has several kinds of symmetry, we expect to clarify the stability of such quasisteady states from the viewpoint of symmetry.

The size dependence of the radiation profile has been investigated in the Dicke model [54]. To clarify the differences between optical near fields and propagating fields, we should also examine such a size dependence in our model. The semiclassical approach might be useful for such a qualitative study because it has an advantage over the quantum approach with respect to computation time.

One of our main goals was to clarify the inherent characteristics in optical near fields from the viewpoint of the coupling scheme, i.e., a local coupling system versus a global coupling system. We have shown several differences between our model (local coupling system) and the Dicke model (global coupling system), but further effort is required to answer the question more directly. In spite of some common properties, we must emphasize that our model differs from the Frenkel exciton model of Tokihiro et al., which is an extended Dicke model that considers the dipole–dipole interaction between excitons. The similarities of the radiation properties in our model and the Tokihiro model arise from the similar master equations used (see (135) and (160)). However, our master equation (160) was derived after neglecting the product of excitons with an order higher than two (see (159)). This approximation is justified for a dilute excitonic system. Nevertheless, the precise conditions under which this approximation is applicable and its consistency should be investigated thoroughly.

Therefore, our localized photon model concurs with the model of Tokihiro et al. only in the simple case, and the true dynamics in our model is more complicated. Unfortunately, there is no straightforward way to derive a more correct equation beyond the approximation (159). The difficulty comes from how to treat the dynamics of three different degrees of freedom: excitons, localized photons, and radiation photons. Rigorously speaking, the difficulty arises from the fact that the unperturbed part $H = H_{\mathrm{a}}+H_{\mathrm{b}}+H_{\mathrm{int}}$ (4) is made up of a mixture of spin operators (exitons) and bosons (localized photons). However, it is possible to consider the next order of excitons in U_{QDeff} in (157) beyond the approximation (159). To proceed to higher orders, we need to remove the time integration in (157), and may use the time-convolutionless (TCL) formulation of the projection operator method [90, 91].

Finally, we focused on the nonlinearity of the Heisenberg equations in our system, and investigated the transport phenomena of localized photons, adding a reservoir of localized photons to the system. Neither normal macroscopic transport phenomena nor equilibrium states were found, which indicates that the nonlinearity of this system is very weak, as a fully nonlinear system, such as a chaotic system, shows statistical transport phenomena (e.g., diffusion) [87, 92] and large-scale simulations are required to find the equilibrium states. Instead, as a transient behavior, we found a remarkable phenomenon in which the transfer of localized photons is jammed or through flowing, depending on the system parameters as ε, V, E, and U. We called the former case the storage mode of localized photons. Moreover, the number of localized photons stored varies drastically, depending on the localized photon–exciton coupling constant $U = \hbar g$. The dependence of the total number of localized photons on the coupling constant g is chaotic and unpredictable. Reduction of the system to the minimal size did not change this behavior; therefore, the complex behavior arose from the intrinsic nonlinearity of the system, not from the interference of many modes due to many degrees of freedom. Since the unpredictable coupling-constant dependence had both regular windows and chaotic regions (see Fig. 43), our system shows "intermittent chaos" in terms of usual dynamical chaos. We found that the chaotic regions have similar distributions, but the meanings of the distributions are not clear. We also checked the dynamics using the next-order quantum correlation and observed the storage and through-flowing modes. The semiclassical treatment seems sufficient to produce these phenomena.

In summary, we found an interesting phenomenon arising from the nonlinearity in our system. The nonlinearity of the dynamics originates from the commutation relation of excitons that differs from that of bosons. This is true in the Dicke model. Therefore, the Dicke model has the same nonlinearity, which is essential for superradiance [58], and certain kinds of signatures of chaos. Indeed, it was recently reported that the Dicke model is a quantum chaotic system [93]. The term "quantum chaotic system" means that the distribution of the energy-level spacing of the Hamiltonian differs from that of a regular system [88]. Since there is no instability of orbits in a quantum system, due to the uncertainty principle, we need new signatures of chaos for a quantum system whose classical correspondence is also chaotic. One of the quantum signatures of chaos is the level statistics [89], and the analysis of these level statistics remains an intriguing open problem. Since our system has a richer structure than the Dicke model, we expect our model to produce more fruitful properties as a quantum chaotic system. Finally, we point out that the open system discussed using our localized photon model is one of the simplest nano-optical structures to use to examine the optical response of nanostructures and to discuss the susceptibilities of nanometric material systems.

Acknowledgements

The authors are grateful to H. Hori, I. Banno, N. Ohnishi, and C. Uchiyama of Yamanashi University, T. Kawazoe of the Japan Science and Technology Agency, and H. Nejo of the National Institute for Materials Science for fruitful discussions. They also gratefully acknowledge valuable suggestions from F. Shibata of Ochanomizu University on the topics in Sect. 5.1.

References

1. E.A. Synge: Philos. Mag. **6**, 356 (1928)
2. M. Ohtsu, H. Hori: *Near-Field Nano-Optics* (Kluwer Academic/Plenum publishers, New York 1999)
3. M. Ohtsu (ed.): *Near-Field Nano/Atom Optics and Technology* (Springer-Verlag, Tokyo 1998)
4. D.W. Pohl, D. Courjon: *Near Field Optics* (Kluwer Academic Publishers, Dordrecht 1993)
5. J.P. Fillard: *Near Field Optics and Nanoscopy* (World Scientific, Singapore 1996)
6. M.A. Paesler, P.J. Moyer: *Near-Field Optics: Theory, Instrumentation, and Application* (Wiley, New York 1996)
7. M. Ohtsu (ed.): *Progress in Nano Electro-Optics I* (Springer-Verlag, Berlin 2002)
8. M. Ohtsu (ed.): *Progress in Nano Electro-Optics II* (Springer-Verlag, Berlin 2003)
9. M. Ohtsu, K. Kobayashi: *Optical Near Fields* (Springer-Verlag, Heidelberg 2004)
10. K. Shimoda: *Introduction to Laser Physics* (Springer-Verlag, Berlin and Heidelberg 1986)
11. C. Carniglia, L. Mandel: Phys. Rev. D **3**, 280 (1971)

12. I. Bialynicki-Birula, J. B. Brojan: Phys. Rev. D **5**, 485 (1972)
13. R. Asby, E. Wolf: J. Opt. Soc. Am. **61**, 52 (1971)
14. E. Lalor, E. Wolf: Phys. Rev. Lett. **26**, 1274 (1971)
15. O. Costa de Beauregard, C. Imvert, J. Ricard: Int. J. Theor. Phy. **4**, 125 (1971)
16. J.M. Vigoureux, R. Payen: J. de Phys. **35**, 617 (1974)
17. J.M. Vigoureux, R. Payen: J. de Phys. **36**, 631 (1975)
18. J.M. Vigoureux, R. Payen: J. de Phys. **36**, 1327 (1975)
19. J.M. Vigoureux, L. D'Hooge, D. Van Labeke: Phys. Rev. A **21**, 347 (1980)
20. S. Huard: Opt. Commun. **30**, 8 (1979)
21. J.M. Vigoureux: 'Emission and absorption of light by electrons or atoms in optical near fields'. In *Near Field Optics* ed. by D.W. Pohl, D. Courjon, NATO ASI Series **242** (Kluwer Academic Publishers, Dordrecht 1993), pp. 239–246
22. M. Janowicz, W. Zakowicz: Phys. Rev. A **50**, 4350 (1994)
23. T. Inoue, H. Hori: Phys. Rev. A **63**, 063805 (2001)
24. H. Khosravi, R. Loudon: Proc. Roy. Soc. London A **433**, 337 (1991)
25. Y. Ohdaira, K. Kijima, K. Terasawa, M. Kawai, H. Hori, K. Kitahara: J. Micros. **202**, 255 (2001)
26. H. Hori, K. Kitahara,, M. Ohtsu: In *Abstracts of the 1st AP-NFO*, 49 (1996)
27. A. Shojiguchi, K. Kitahara, K. Kobayashi, M. Ohtsu: J. Micros. **210**, 301 (2003)
28. H. Levine, J. Schwinger: Phys. Rev. **74**, 958 (1948)
29. H.A. Bethe: Phys. Rev. **66**, 163 (1944)
30. C.J. Bouwkamp: Rep. Prog. Phys. **17**, 35 (1954)
31. C.J. Bouwkamp: Philips Res. Rep. **5**, 321 (1950)
32. C.J. Bouwkamp: Philips Res. Rep. **5**, 401 (1950)
33. T.H. Stievater, Xiaoqin Li, D.G. Steel, D. Gammon, D.S. Katzer, D. Park, C. Piermarocchi, L.J. Sham: Phys. Rev. Lett. **87**, 133603 (2001)
34. H. Kamada, H. Gotoh, J. Temmyo, T. Takagahara, H. Ando: Phys. Rev. Lett. **87**, 246401 (2001)
35. H. Htoon, T. Takagahara, D. Kulik, O. Baklenov, A.L. Holmes, Jr., C.K. Shih: Phys. Rev. Lett. **88**, 087401 (2002)
36. T. Kawazoe, K. Kobayashi, J. Lim, Y. Narita, M. Ohtsu: Phys. Rev. Lett. **88**, 067404 (2002)
37. M. Ohtsu, K. Kobayashi, T. Kawazoe, S. Sangu, T. Yatsui: IEEE J. Selec. Topics Quantum Electron. **8**, 839 (2002)
38. K. Cho: Suppl. Prog. Theor. Phys. **106**, 225 (1991)
39. H. Ishihara, K. Cho: Phys. Rev. B **48**, 7960 (1993)
40. K. Cho, Y. Ohfuti, K. Arima: Surf. Sci. **363**, 378 (1996)
41. R.T. Cox: Phys. Rev. **66**, 106 (1944)
42. J.G. Linhart: J. Appl. Phys. **26**, 527 (1955)
43. K. Kobayashi, M. Ohtsu: J. Micros. **194**, 249 (1999)
44. K. Kobayashi, S. Sangu, H. Ito, M. Ohtsu: Phys. Rev. A **63**, 013806 (2001)
45. K. Kobayashi, S. Sangu, M. Ohtsu: 'Quantum theoretical approach to optical near-fields and some related applications'. In *Progress in Nano-Electro-Optics I* ed. by M. Ohtsu, Springer Series in Optical Sciences (Springer-Verlag, Berlin, 2002) pp. 119–157
46. A. Shojiguchi, K. Kobayashi, S. Sangu, K. Kitahara, M. Ohtsu: Nonlin. Opt. **29**, 563 (2002)
47. S. Sangu, K. Kobayashi, A. Shojiguchi, T. Kawazoe, M. Ohtsu: J. Appl. Phys. **93**, 2937 (2003)

48. K. Kobayashi, S. Sangu, A. Shojiguchi, T. Kawazoe, K. Kitahara, M. Ohtsu: J. Micros. **210**, 247 (2003)
49. R.H. Dicke: Phys. Rev. **93**, 99 (1954)
50. L. Quiroga, N.F. Johnson: Phys. Rev. Lett. **83**, 2270 (1999)
51. H.A. Posch, W.G. Hoover, F.J. Vesely: Phys. Rev. A **33**, 4253 (1986)
52. W.G. Hoover, H.A. Posch: Phys. Rev. E **49**, 1913 (1994)
53. A. Shojiguchi, K. Kobayashi, S. Sangu, K. Kitahara, M. Ohtsu: J. Phys. Soc. Jpn. **72**, 2984 (2003)
54. M. Gross, S. Haroche: Phys. Rep. **93**, 301 (1982)
55. N. Rehler, J.H. Eberly: Phys. Rev. A **3**, 1735 (1971)
56. R.H. Lehmberg: Phys. Rev. A **2**, 883 (1970)
57. G.S. Agarwal: Phys. Rev. A **2**, 2038 (1970)
58. G.S. Agarwal: Phys. Rev. A **4**, 1791 (1971)
59. L.M. Narducci, C.A. Coulter, C.M. Bowden: Phys. Rev. A **9**, 829 (1974)
60. R. Bonifacio, P. Schwendimann, F. Haake: Phys. Rev. A **4**, 302 (1971)
61. R. Friedberg, S.R. Hartmann: Phys. Rev. A **10**, 1728 (1974)
62. B. Coffey, R. Friedberg: Phys. Rev. A **17**, 1033 (1978)
63. R. Bonifacio, L.A. Lugiato: Phys. Rev. A **11**, 1507 (1975)
64. N. Skribanowitz, I.P. Herman, J.C. MacGillivray, M.S. Feld: Phys. Rev. Lett. **30**, 309 (1973)
65. R. Bonifacio, L.A. Lugiato, A.A. Crescentini: Phys. Rev. A **13**, 1507 (1976)
66. A. Nakamura, H. Yamada, T. Tokizaki: Phys. Rev. B **40**, 8585 (1989)
67. Y.N. Chen, D.S. Chuu, T. Brandes: Phys. Rev. Lett. **90**, 166801 (2003)
68. N. Piovella, V. Beretta, G.R.M. Robb, R. Bonifacio: Phys. Rev. A **68**, 021801 (2003)
69. T. Tokihiro, Y. Manabe, E. Hanamura: Phys. Rev. B **47**, 2019 (1993)
70. T. Brandes, J. Inoue, A. Shimizu: Phys. Rev. Lett. **80**, 3952 (1998)
71. T. Saso, Y. Suzumura, H. Fukuyama: Suppl. Prog. Theoret. Phys. **84**, 269 (1985)
72. V.M. Agranovich, M.D. Galanin: *Electric Excitation Energy Transfer in Condensed Matter* (North-Holland, Amsterdam 1982)
73. J. Knoester, S. Mukamel: Phys. Rev. A **39**, 1899 (1989)
74. H. Fröhlich: Phys. Rev. **79**, 845 (1950)
75. J. Bardeen, D. Pines: Phys. Rev. **99**, 1140 (1955)
76. C. Kittel: *Quantum Theory of Solids*, 2nd revised printing (John Wiley & Sons, New York 1987)
77. L. Mandel, E. Wolf: *Optical Coherence and Quantum Optics* (Cambridge University Press, Cambridge 1995)
78. G.S. Agarwal: Phys. Rev. **178**, 2025 (1969)
79. S. Nakajima: Prog. Theor. Phys. **20**, 948 (1958)
80. R. Zwanzig: J. Chem. Phys. **33**, 1338 (1960)
81. R. Zwanzig: Phys. Rev. **124**, 983 (1961)
82. R. Kubo, M. Toda, N. Hashitsume: *Statistical Physics II* (Springer-Verlag, Berlin 1985)
83. T. Tokihiro, Y. Manabe, E. Hanamura: Phys. Rev. B **51**, 7655 (1995)
84. E. Lieb, T. Schultz, D. Mattis: Annal. Phys. **16**, 407 (1961)
85. R. Bonifacio, L.A. Lugiato: Phys. Rev. A **12**, 587 (1975)
86. G. Lindblad: Comun. Math. Phys. **48**, 119 (1976)
87. P. Gaspard: *Chaos, Scattering and Statistical Mechanics* (Cambridge University Press, Cambridge 1998)

88. M.C. Gutzwiller: *Chaos in Classical and Quantum Mechanics* (Springer-Verlag, Berlin 1995)
89. F. Haake: *Quantum Signatures of Chaos* (Springer-Verlag, Berlin 1991)
90. S. Chaturvedi, F. Shibata: Z. Phys. B **35**, 297 (1979)
91. F. Shibata, T. Arimitsu: J. Phys. Soc. Jpn. **49**, 891 (1980)
92. D.J. Evans, G.P. Morriss: *Statistical Mechanics of Nonequilibrium Liquids* (Academic Press, New York 1997)
93. C. Emary, T. Brandes: Phys. Rev. Lett. **90**, 044101 (2003)

Index

Springer Series in
OPTICAL SCIENCES

New editions of volumes prior to volume 70

1 **Solid-State Laser Engineering**
By W. Koechner, 5th revised and updated ed. 1999, 472 figs., 55 tabs., XII, 746 pages

14 **Laser Crystals**
Their Physics and Properties
By A. A. Kaminskii, 2nd ed. 1990, 89 figs., 56 tabs., XVI, 456 pages

15 **X-Ray Spectroscopy**
An Introduction
By B. K. Agarwal, 2nd ed. 1991, 239 figs., XV, 419 pages

36 **Transmission Electron Microscopy**
Physics of Image Formation and Microanalysis
By L. Reimer, 4th ed. 1997, 273 figs. XVI, 584 pages

45 **Scanning Electron Microscopy**
Physics of Image Formation and Microanalysis
By L. Reimer, 2nd completely revised and updated ed. 1998, 260 figs., XIV, 527 pages

Published titles since volume 70

70 **Electron Holography**
By A. Tonomura, 2nd, enlarged ed. 1999, 127 figs., XII, 162 pages

71 **Energy-Filtering Transmission Electron Microscopy**
By L. Reimer (Ed.), 1995, 199 figs., XIV, 424 pages

72 **Nonlinear Optical Effects and Materials**
By P. Günter (Ed.), 2000, 174 figs., 43 tabs., XIV, 540 pages

73 **Evanescent Waves**
From Newtonian Optics to Atomic Optics
By F. de Fornel, 2001, 277 figs., XVIII, 268 pages

74 **International Trends in Optics and Photonics**
ICO IV
By T. Asakura (Ed.), 1999, 190 figs., 14 tabs., XX, 426 pages

75 **Advanced Optical Imaging Theory**
By M. Gu, 2000, 93 figs., XII, 214 pages

76 **Holographic Data Storage**
By H.J. Coufal, D. Psaltis, G.T. Sincerbox (Eds.), 2000
228 figs., 64 in color, 12 tabs., XXVI, 486 pages

77 **Solid-State Lasers for Materials Processing**
Fundamental Relations and Technical Realizations
By R. Iffländer, 2001, 230 figs., 73 tabs., XVIII, 350 pages

78 **Holography**
The First 50 Years
By J.-M. Fournier (Ed.), 2001, 266 figs., XII, 460 pages

79 **Mathematical Methods of Quantum Optics**
By R.R. Puri, 2001, 13 figs., XIV, 285 pages

80 **Optical Properties of Photonic Crystals**
By K. Sakoda, 2nd ed., 2004, 107 figs., 29 tabs., XIV, 255 pages

81 **Photonic Analog-to-Digital Conversion**
By B.L. Shoop, 2001, 259 figs., 11 tabs., XIV, 330 pages

82 **Spatial Solitons**
By S. Trillo, W.E. Torruellas (Eds), 2001, 194 figs., 7 tabs., XX, 454 pages

83 **Nonimaging Fresnel Lenses**
Design and Performance of Solar Concentrators
By R. Leutz, A. Suzuki, 2001, 139 figs., 44 tabs., XII, 272 pages

84 **Nano-Optics**
By S. Kawata, M. Ohtsu, M. Irie (Eds.), 2002, 258 figs., 2 tabs., XVI, 321 pages

Springer Series in
OPTICAL SCIENCES

85 **Sensing with Terahertz Radiation**
By D. Mittleman (Ed.), 2003, 207 figs., 14 tabs., XVI, 337 pages

86 **Progress in Nano-Electro-Optics I**
Basics and Theory of Near-Field Optics
By M. Ohtsu (Ed.), 2003, 118 figs., XIV, 161 pages

87 **Optical Imaging and Microscopy**
Techniques and Advanced Systems
By P. Török, F.-J. Kao (Eds.), 2003, 260 figs., XVII, 395 pages

88 **Optical Interference Coatings**
By N. Kaiser, H.K. Pulker (Eds.), 2003, 203 figs., 50 tabs., XVI, 504 pages

89 **Progress in Nano-Electro-Optics II**
Novel Devices and Atom Manipulation
By M. Ohtsu (Ed.), 2003, 115 figs., XIII, 188 pages

90/1 **Raman Amplifiers for Telecommunications 1**
Physical Principles
By M.N. Islam (Ed.), 2004, 488 figs., XXVIII, 328 pages

90/2 **Raman Amplifiers for Telecommunications 2**
Sub-Systems and Systems
By M.N. Islam (Ed.), 2004, 278 figs., XXVIII, 420 pages

91 **Optical Super Resolution**
By Z. Zalevsky, D. Mendlovic, 2004, 164 figs., XVIII, 232 pages

92 **UV-Visible Reflection Spectroscopy of Liquids**
By J.A. Räty, K.-E. Peiponen, T. Asakura, 2004, 131 figs., XII, 219 pages

93 **Fundamentals of Semiconductor Lasers**
By T. Numai, 2004, 166 figs., XII, 264 pages

94 **Photonic Crystals**
Physics, Fabrication and Applications
By K. Inoue, K. Ohtaka (Eds.), 2004, 209 figs., XV, 320 pages

95 **Ultrafast Optics IV**
Selected Contributions to the 4th International Conference
on Ultrafast Optics, Vienna, Austria
By F. Krausz, G. Korn, P. Corkum, I.A. Walmsley (Eds.), 2004, 281 figs., XIV, 506 pages

96 **Progress in Nano-Electro Optics III**
Industrial Applications and Dynamics of the Nano-Optical System
By M. Ohtsu (Ed.), 2004, 186 figs., 8 tabs., XIV, 224 pages

97 **Microoptics**
From Technology to Applications
By J. Jahns, K.-H. Brenner, 2004, 303 figs., XI, 335 pages

98 **X-Ray Optics**
High-Energy-Resolution Applications
By Y. Shvyd'ko, 2004, 181 figs., XIV, 404 pages

99 **Few-Cycle Photonics and Optical Scanning Tunneling Microscopy**
Route to Femtosecond Ångstrom Technology
By M. Yamashita, H. Shigekawa, R. Morita (Eds.) 2005, 241 figs., XX, 393 pages

Printing: Saladruck, Berlin
Binding: Stein+Lehmann, Berlin